Let's
Botanize!

Let's Botanize!

101 Ways to Connect with Plants

Find

Look

Compare

Ben Goulet-Scott
Jacob S. Suissa

Storey Publishing

The mission of Storey Publishing is to serve our customers by publishing practical information that encourages personal independence in harmony with the environment.

EDITED BY Hannah Fries

ART DIRECTION AND BOOK DESIGN BY Carolyn Eckert

TEXT PRODUCTION BY Jennifer Jepson Smith

COVER PHOTOGRAPHY BY © Let's Botanize, Inc. except David Clode/Unsplash, front; Rlevse/CC BY-SA 3.0/Wikimedia Commons, back m.

INTERIOR PHOTOGRAPHY BY © Let's Botanize, Inc.

ADDITIONAL INTERIOR PHOTOGRAPHY CREDITS appear on page 239.

ILLUSTRATIONS BY © Nina Chakrabarti

Storey books may be purchased in bulk for business, educational, or promotional use. Special editions or book excerpts can also be created to specification. For details, please contact your local bookseller or the Hachette Book Group Special Markets Department at special.markets@hbgusa.com.

Storey Publishing
210 MASS MoCA Way
North Adams, MA 01247
storey.com

Storey Publishing is an imprint of Workman Publishing, a division of Hachette Book Group, Inc., 1290 Avenue of the Americas, New York, NY 10104. The Storey Publishing name and logo are registered trademarks of Hachette Book Group, Inc.

ISBNs: 978-1-63586-904-0 (paperback); 978-1-63586-905-7 (ebook)

Printed in China by Toppan Leefung Printing Ltd. on paper from responsible sources
10 9 8 7 6 5 4 3 2 1

TLF

Library of Congress Cataloging-in-Publication Data on file

What's Botanizing?

Botany is the study of plants. To turn the word from an academic category into an active pastime—like birding—we simply add an -*ize* suffix. To *botanize* is to spend time alongside plants with the specific intention of observing and appreciating them as living organisms.

Botanizing is a way to enrich your existing relationship with plant life. Our lives are inexorably linked with the lives of plants, and there's a good chance you're more familiar with them than you realize. With more than 350,000 living species of land plants, you'd be hard-pressed to go anywhere and *not* be able to closely observe plant life.

Botanizing is learning by exploring. It's hands-on, it's immersive, and it's very easy to get started. You can botanize almost anywhere. Plants don't just exist in wilderness areas; they fill our cities, our parks, our gardens, and our kitchens. In fact, an estimated 82 percent of the living stuff (biomass) on land is plant life. Plants form the foundation of nearly every ecosystem on Earth. Importantly, they do not move, so you won't have to chase them down, and you can get as close as you like.

There are more species of grass than birds,
more mints than mammals,
and more beans
than butterflies.

While learning to identify different species of plants adds a layer of intimacy to interacting with these amazing organisms, you do not need to memorize Latin names or dichotomous keys to start botanizing. In fact, you don't need to focus on identification at all, if you don't want to.

The core of botanizing is carefully observing plant form and function and allowing those observations to drive your curiosity about how plants build their intricate and beautiful bodies, how they interact with other species, and how they survive, and have survived, in the landscape over millennia.

Given that plants form the biological foundation for life on Earth, learning about them is crucial for understanding how life on this planet evolved and is sustained. And yet we often take plants for granted. Ironically, it may be exactly because they are so ubiquitous that we look right past them. Plants are woven so integrally into nearly every aspect of our daily lives that their presence feels like a default setting. We didn't turn this setting on, and we can't imagine turning it off.

Therefore, it can be easy to mistake plant life as little more than the stage on which our (animal) existence plays out.

But that green substrate of our lives encompasses an astounding diversity of form and function, reflecting hundreds of millions of years of evolutionary history.

From this diversity, plants provide us with the raw materials for the homes we live in, the food we eat, and the oxygen we breathe. Increasingly, societies are recognizing the physical and mental health benefits of spending time in green, plant-filled spaces. Botanizing leads us into green spaces to engage with our local human and nonhuman communities. These experiences allow us to connect more deeply with plants as organisms.

Ultimately, caring for plants is caring for our planet.

Whether you are just getting started or are already an experienced plant observer, this book is our attempt to inspire you to embark on a deeper journey into the world of plants.

In this book, you will find 101 prompts to resensitize your attention to plants and activate your botanical curiosity. We hope you find joy in them, too. They are organized into three sections:

Parts is concerned with the different elements that make up a plant's body (such as stems, roots, leaves, and flowers). These prompts start on a small scale, focusing on plant structure, anatomy, morphology, and organization.

Patterns builds on the foundation of the first section and focuses on processes of plant survival. These prompts explore plant development, physiology, evolution, and ecology to investigate how plants build their bodies and how they function in intricate and deeply interdependent ecosystems.

Perspectives focuses on mindful engagement with plants at different scales of time and space. These prompts ask us to interact with plants as living organisms and to appreciate them as individuals that are wholly distinct from us but possible to connect with. They are not like us, and that is why they are beautiful and fascinating.

Prompts can be done in the order they are numbered, building in scale from specific to grand, or they can be explored in any order you wish. They can be done alone or with others. You can botanize as the primary goal on your outings, or you can do it in tandem with other nature-based activities like rock climbing or bird-watching—or just on the way to the store. Botanizing can be a reflective and deeply personal experience, or it can be a highly social pastime.

As scientists, we wrote these prompts with the intention to never compromise accuracy or simplify concepts for convenience. We also aimed to make the content accessible by eliminating or clearly defining jargon and botanical terminology. The glossary at the back of the book will not only help you understand the prompts but will assist you in your botanical journey.

This collection of prompts will never expire. It can be revisited in different places, in different seasons, and at different levels of experience, with ever more to notice and enjoy. The botanizing game never ends. Finally, there is no test at the end of this book, and not all questions have answers. By no means can the full breadth of the botanical world be covered in just 101 short prompts—this is only an entry into the world of these marvelous organisms. There is so much our species does not know about plants. Allow mystery to drive your curiosity. Explore for the joy of learning, challenge yourself, have fun, and seek wonder in the green.

Let's Botanize!
Ben Goulet-Scott & Jacob S. Suissa

The 5 Tenets of Botanizing

Botanize any time, any place.

Plants, wild or cultivated, and plant parts are around you every day, no matter where you are. The forest, sidewalk, and kitchen are all exciting places to botanize.

Get as close as possible.

Plants don't move, and this allows us to access the intricacy of their minute details. Use a hand lens or magnifying glass to observe their minutiae.

Engage all of your senses.

Plant variation is chemical as well as physical; observe with more than just your eyes.

Identification is not the endgame.

Identification is a fun challenge and can help elevate the intimate relationships you build with plants, but the core of botanizing is observing plant form and function.

Rediscover the familiar.

There is a special joy that comes from noticing new details in familiar plants, and there's always something new to discover.

parts

1
Look closely at the patterns of leaf veins of two different species.

Just like insects, birds, and humans, plants have vascular systems—sets of tubes that move water and nutrients through their bodies. These veins are found all throughout a plant, in its roots, stems, and leaves.

The cells that make up veins are composed of specific tissues: **xylem** (water-conducting tissue) and **phloem** (sugar-conducting tissue). Leaf veins are particularly important because they are the interface between the plant and the atmosphere. At the smallest veinlets, water leaks out of the water-conducting cells to supply the leaf tissue with vital moisture; at the same time, much of the water that was sucked up from the soil evaporates into the air through tiny pores in the leaf called **stomata**.

Leaf veins display diverse patterning across the plant tree of life, from single veins to complex networks, each pattern as distinct as a fingerprint and with a story to tell.

The ginkgo (*Ginkgo biloba*) has simple, **dichotomous** veins that branch two at a time, which is rare in **seed** plants but more common in ferns.

This catalpa leaf (*Catalpa speciosa*) has a large primary vein with many smaller branching veins in between. A netlike pattern of veins is described as **reticulate.**

If you look closely at a leaf, you will see a repeated or fractal vein pattern.

WHEN DO LEAF VEINS DEVELOP?

Leaf veins are laid down early in the life of a leaf, when it is less than a centimeter in length, sitting inside a **bud**. The patterning of these veins is determined by the spread of hormones through a developing leaf, which signals to cells whether they should turn into **vascular tissues**. This process is fine-tuned to distribute veins across the full area of the leaf (see prompt 23).

2
Cut open two fruits and compare their contents.

Do they have one seed or many?
Is the fruit fleshy or hard?

Fruits come from flowers.

In fact, they are part of the flower (see prompt 38). A **fruit** is the plant's **ovary**, which holds the seeds, protects the seeds during development, and helps them disperse when they are mature. True fruits are unique to flowering plants (**angio-sperms**, literally translating from Latin to "closed seed," in reference to the seed being enclosed in the fruit). This is what makes the flowering plants unique— no other plant lineage produces fruit.

Fruits are made of structures called **carpels**, which are the smaller subunits of the ovary. Fruits can be made up of multiple carpels (like an orange) or a single carpel (like a peach), and each carpel can have a single seed or many seeds. A fruit can be fleshy like a blueberry or dry like an acorn. It can be winged like a maple fruit and dispersed by the wind, or juicy like a kiwi and dispersed by animals (see prompt 53).

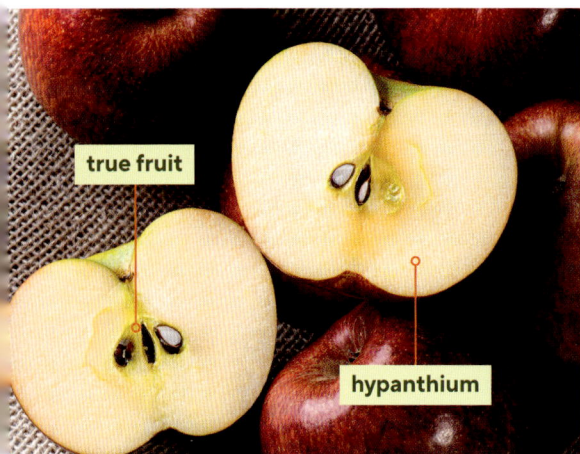

true fruit

hypanthium

IS IT A TRUE FRUIT?

In some cases, the things we call fruits are not fruits at all! For instance, the fleshy part of a strawberry is a swollen stem (the fruits are the small things that get stuck in your teeth), and apples are mostly derived from stemlike tissue called a **hypanthium**; the true fruit of the apple is the papery core that encases the seeds!

parts

The edible **arils** that surround the seeds of this passionflower (*Passiflora coloranigra*) are contained within a brightly colored fruit that will attract animals to eat and disperse it.

endocarp

exocarp

mesocarp

The carpel has three distinct layers of tissue. The outer layer (**exocarp**), middle layer (**mesocarp**), and inner layer (**endocarp**) can form different structures within a fruit. For instance, plums (*Prunus domestica*) often have a waxy exocarp, juicy mesocarp, and hard, stony endocarp.

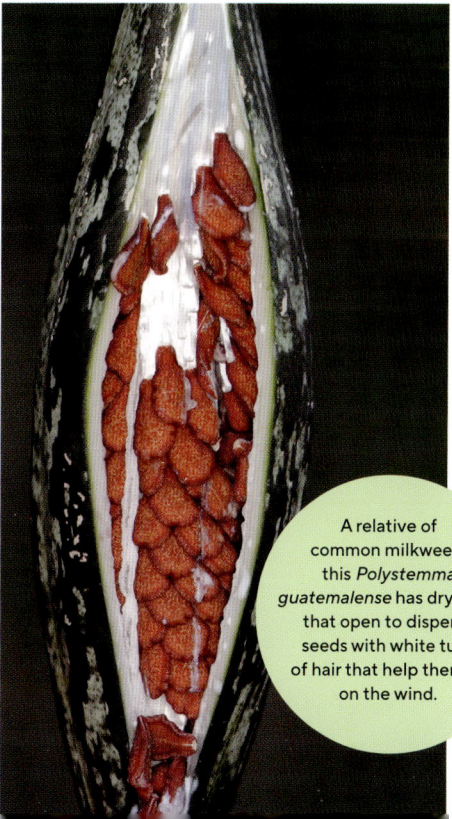

A relative of common milkweed, this *Polystemma guatemalense* has dry fruits that open to disperse seeds with white tufts of hair that help them fly on the wind.

3
Find a bud at the tip of a branch.

Can you see tiny leaves? Is the bud covered by scales, hairs, or something else?

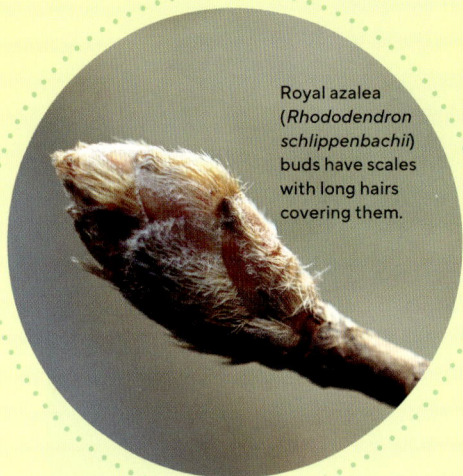

Royal azalea (*Rhododendron schlippenbachii*) buds have scales with long hairs covering them.

Compared to plants, animals are developmentally static organisms. That might sound odd at first. Yes, animals often run, crawl, swim, or fly, while plants are stuck in place. But the animal body fully develops into its mature state and stops. There are some animals that continue to grow—fish, for example—but they do not continue developing; they just expand in size. In contrast, plant development never ends.

A mature tree at 5 years old will look drastically different from that same individual at 20, 50, and 100 years old (see prompt 73). All this aboveground growth originates in the growing point of the plant called the **shoot apical meristem** (SAM for short). This is a vital region of the plant that houses a cluster of the stem cells that give rise to the new organs of the plant body (stems, leaves, and flowers). Many trees and long-lived perennial plants protect their SAMs in buds. Buds can be covered in scales, hairs, or even leaves.

New growth of the northern wild raisin (*Viburnum cassinoides*) is surrounded by a pair of elongated leaves. Buds without scales are often called **naked buds**.

parts

20

The bud scales
of a linden tree
(*Tilia maximowicziana*)
open to reveal new
expanding leaves.

WHAT WE REALLY KNOW ABOUT BUD SCALES

Buds with scales are more prevalent in the temperate region, while most tropical species tend to have naked buds. Bud scales have long been hypothesized to protect the SAM from frost damage. However, there are at least 42 families of temperate plants with naked buds, challenging this hypothesis. While buds may provide some protective function, their structure may also relate to other factors like leaf expansion and longevity during the growing season.

4
Compare the textures and colors of the upper and lower surfaces of a leaf.

Is one side smoother or hairier?

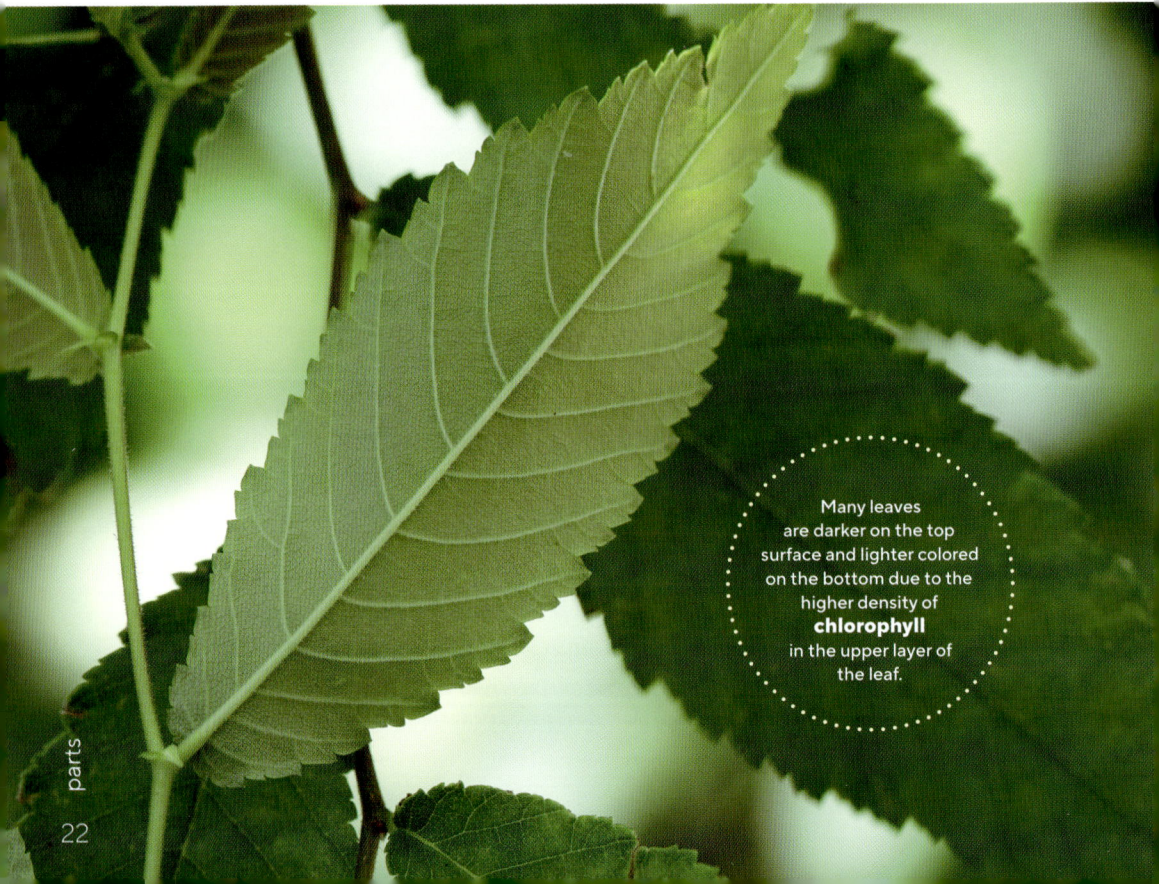

Many leaves are darker on the top surface and lighter colored on the bottom due to the higher density of **chlorophyll** in the upper layer of the leaf.

Most leaves appear to be simple structures—a flat blade of tissue crisscrossed by a network of veins. In reality, leaves are sophisticated contraptions in an elegant package that have evolved to convert energy from the sun into a biologically useful form (sugar) through **photosynthesis**. You can think of a leaf like a sandwich made up of a few different layers of tissue. The upper half of the sandwich specializes in harvesting light and performing photosynthesis, while the lower half specializes in regulating gas and water exchange.

To help prevent them from drying out, many species, like this Makino rhododendron (*Rhododendron makinoi*), have modifications like dense hairs on the undersides of their leaves that may help slow evaporation when stomata are open.

Layers of a Leaf

Cuticle.
This waxy layer secreted by the epidermis provides some protection and prevents the leaf from losing too much water.

Upper epidermis.
The epidermis is kind of like the skin of a leaf.

Palisade layer.
Most photosynthesizing happens here. These cells are packed with **chloroplasts**.

Spongy layer.
The cells in the spongy layer are interspersed with air spaces, which allow carbon dioxide to reach the chloroplasts in the palisade layer above.

Lower epidermis.
This layer contains most of the stomata, pores that open and close to allow carbon dioxide to enter the leaf (and water to exit).

5
Can you find the growing tip of a root?

A long root grows deep into the soil.

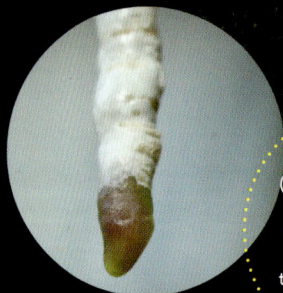

Moth orchids (*Phalaenopsis*) often grow roots that are exposed to the air, allowing us to get a good look at the root tip.

Root Tip Pushing Through the Soil

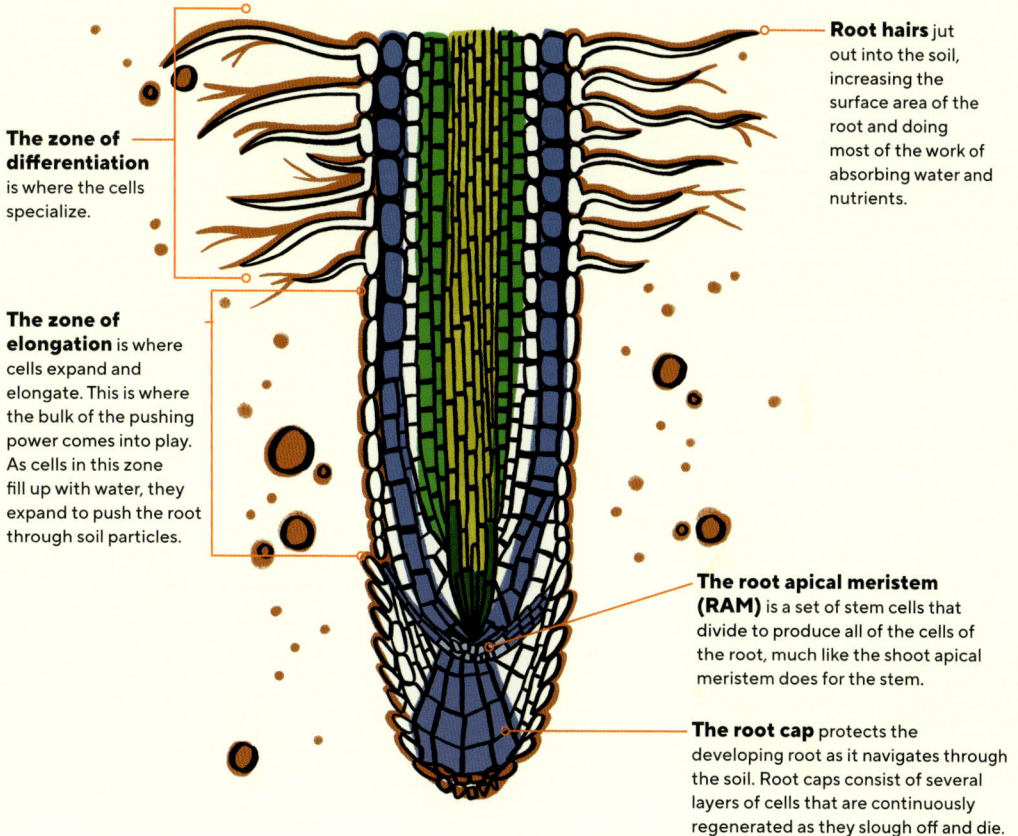

Root hairs jut out into the soil, increasing the surface area of the root and doing most of the work of absorbing water and nutrients.

The zone of differentiation is where the cells specialize.

The zone of elongation is where cells expand and elongate. This is where the bulk of the pushing power comes into play. As cells in this zone fill up with water, they expand to push the root through soil particles.

The root apical meristem (RAM) is a set of stem cells that divide to produce all of the cells of the root, much like the shoot apical meristem does for the stem.

The root cap protects the developing root as it navigates through the soil. Root caps consist of several layers of cells that are continuously regenerated as they slough off and die.

The most obvious parts of a plant are the aboveground organs: stems, leaves, flowers, and fruits. However, there is a whole world belowground. The **shoot system** of a plant (the aboveground parts) is negatively **geotropic**, meaning it grows upward against the direction of gravity. The **root system** is positively geotropic, growing downward in the direction of gravity (see prompt 19 for exceptions).

The structure of the root itself relates to its function in marvelous ways. If you repot your houseplants or pluck up a tiny seedling, you will notice that the tips of the youngest roots are pale or white. In that tiny region are all the cells that give rise to the root and all other roots that come from it. Roots burrow their way through the soil by dividing and expanding the cells produced at their growing tip. Roots also tend to have a phenology, much like leaves, flowers, and fruits: They are often most active early in the growing season.

parts

6

Compare a fern leaf that is visually complex to another that is simpler.

"Nature made ferns for pure leaves, to show what she could do in that line."

This unpublished quote from Henry David Thoreau (documented by Ralph Waldo Emerson in 1862) beautifully conveys how diverse fern leaves can be. Leaves are the most obvious organ of the fern, and they come in many shapes and sizes. But what probably comes to mind when you conjure an image of a fern leaf is a featherlike frond. Indeed, many species have fronds like this. They are called pinnate. Fern fronds are made up of distinct parts: The stem is called the stipe, and the leaf is called the lamina, which can be divided into many leaflets.

In different species of ferns, these organs are modified—lengthened, divided (dissected), and ornamented—resulting in some of the most visually diverse structures in the plant kingdom. They can also be simple leaves with no dissection at all, often resembling leaves of angiosperms. Most simple-leaved ferns are tropical, but you may even have one growing as a houseplant.

A northern maidenhair fern (*Adiantum pedatum*) has a single leaf with a once-divided (sometimes called pedate) leaf form.

This *Pleopeltis panamensis* has a simple, undivided leaf. Spores are produced along the underside.

Hay-scented fern (*Sitobolium punctilobulum*) is twice or even three times divided. The **pinnae** (leaflets) and **costae** (stalks of the leaflets) are divided into what are called **pinnules** and **costules**.

Parts of a Fern Leaf

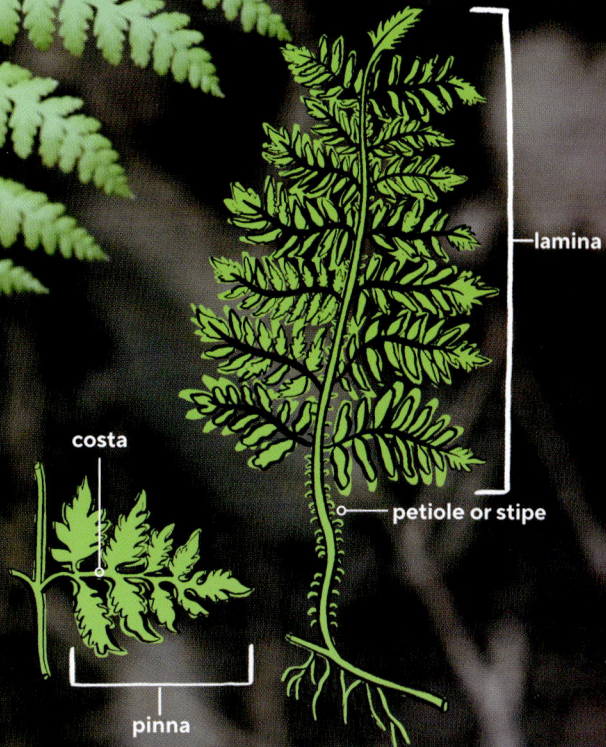

lamina

petiole or stipe

costa

pinna

The titan arum (*Amorphophallus titanum*) looks like the largest flower in the world, but it's not! Like other plants in the Arum family, its flowers are tiny. They are hidden by the hundreds beneath the purple outer sheath called the spathe.

7
What are the smallest and largest flowers you can find?

Finding the limits of nature has always fascinated humans.

We constantly wonder what the oldest tree is, the fastest mammal, the largest plant, and the smallest insect. Indeed, this helps us get a handle on the vast diversity of life. So, how large or small can a flower be? Turns out the variation can be immense.

To our knowledge, the plants with the smallest flowers in the world are duck-weeds in the genus *Wolffia*. This minute aquatic plant also has the smallest leaves in the world. The largest flower in the world is the stinking corpse lily, *Rafflesia arnoldii*. The size difference between *Wolffia* and *Rafflesia arnoldii* is nearly four orders of magnitude—which means the corpse flower is nearly 10,000 times larger!

Duckweed (genus *Wolffia*) is in the same family as titan arum, the Arum family. Its single flower (not visible here) is barely wider than the thickness of paper.

Rafflesia is a parasite. It does not produce leaves, roots, or stems. Its seeds germinate on the roots of a vine in the grape family (Vitaceae) and grow a rootlike structure called a **haustorium**, which taps into the host plant's tissues to steal nutrients. Once mature, it will send out a flower bud the size of a volleyball that will eventually open into a single flower larger than 3 feet (1 meter) in diameter that weighs over 20 pounds. The thick **petals** smell like rotting meat to attract flies and beetles that help pollinate it.

parts

Tuberous grasspink (*Calopogon tuberosus*) is a zygomorphic flower pollinated by bees.

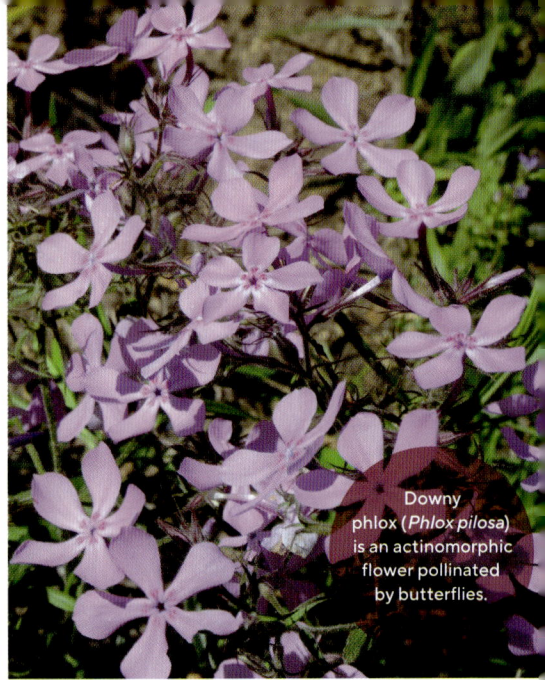

Downy phlox (*Phlox pilosa*) is an actinomorphic flower pollinated by butterflies.

8

Can you find two flowers with different patterns of symmetry?

With over 300,000 flowering plant species, the diversity of flower form is immense.
Some are radially symmetrical, called **actinomorphic** (think starfish). Others are bilaterally symmetrical, called **zygomorphic** (think lobster). Flowers have evolved different shapes and colors to attract animal pollinators, so when we look at a flower, we can often use its shape and colors as clues to predict what animal pollinates it. For instance, actinomorphic flowers that are yellow or white are generally pollinated by insects like bees, while tubular zygomorphic flowers that are red, orange, or white may be pollinated by hummingbirds, butterflies, or moths.

While these general patterns hold true, these animals and plants do not necessarily conform to the boxes we place around them. If there is food (nectar or **pollen**) to be had in a flower, rest assured that an animal will seek it out (see prompt 52).

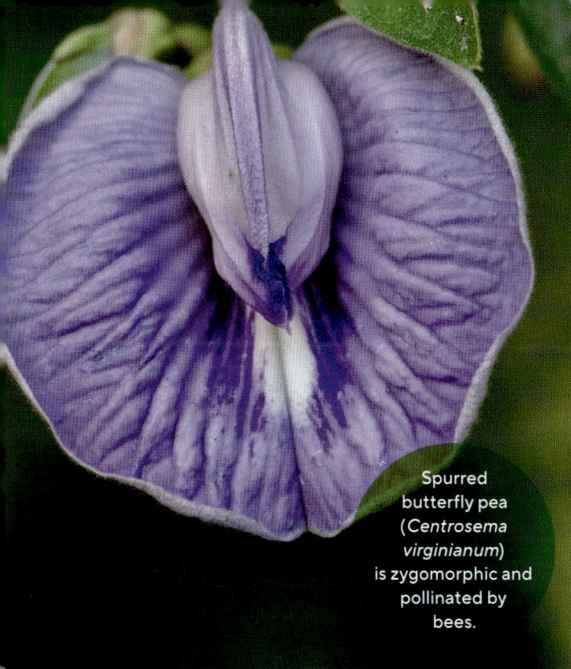

Spurred butterfly pea (*Centrosema virginianum*) is zygomorphic and pollinated by bees.

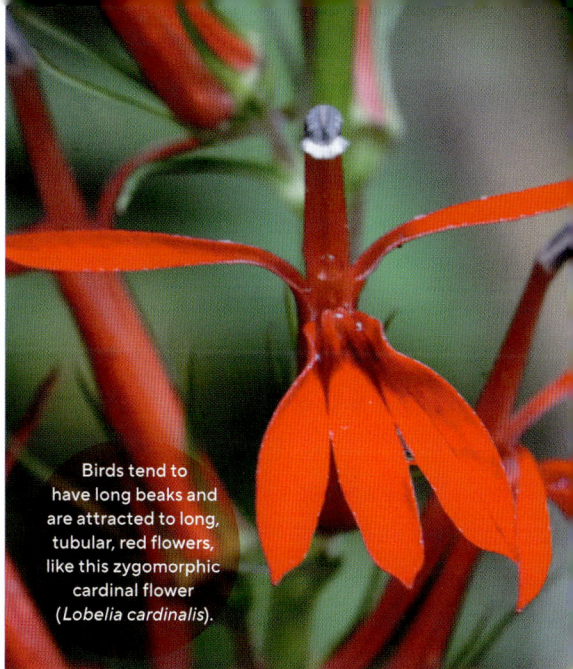

Birds tend to have long beaks and are attracted to long, tubular, red flowers, like this zygomorphic cardinal flower (*Lobelia cardinalis*).

Bees see color in the UV spectrum and can be attracted to large, open-faced yellow flowers, like this actinomorphic Jerusalem artichoke (*Helianthus tuberosus*).

9

Compare a flower with fused petals to one with separate petals.

Regardless of the size and shape of a flower, all are composed of some or all of the same parts. These parts include the **sepals**, petals, **stamens**, and carpels. Some species have evolved flowers that are missing one or more of these components or produce two types of flowers, each with a subset of the four core floral parts.

One way flowering plants achieve the vast diversity of floral shapes is through fusing or separating their sepals, petals, stamens, and carpels. When floral organs are fused, they are **connate**. Connate petals produce a floral tube called a **corolla**. Because they can be more difficult to enter, tubular flowers are often associated with specialized pollinators.

A KEY TO DIVERSITY

If some individuals within a species develop fused flowers and pollinator specialization, over time they may become a distinct species. In fact, floral fusion is one of the main traits associated with the diversification of flowering plants throughout the world. And this is just fusion of the same organ types. Whole new worlds of diversity are possible when organs are **adnate**, fused to unlike organs— for instance, stamens fused to petals, or petals fused to sepals.

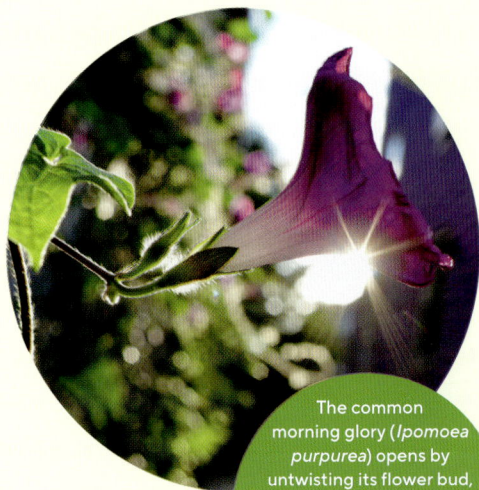

The common morning glory (*Ipomoea purpurea*) opens by untwisting its flower bud, leaving fold lines imprinted on its fused petals. It opens once in the morning, and, upon closure, will not reopen.

Parts of a Flower

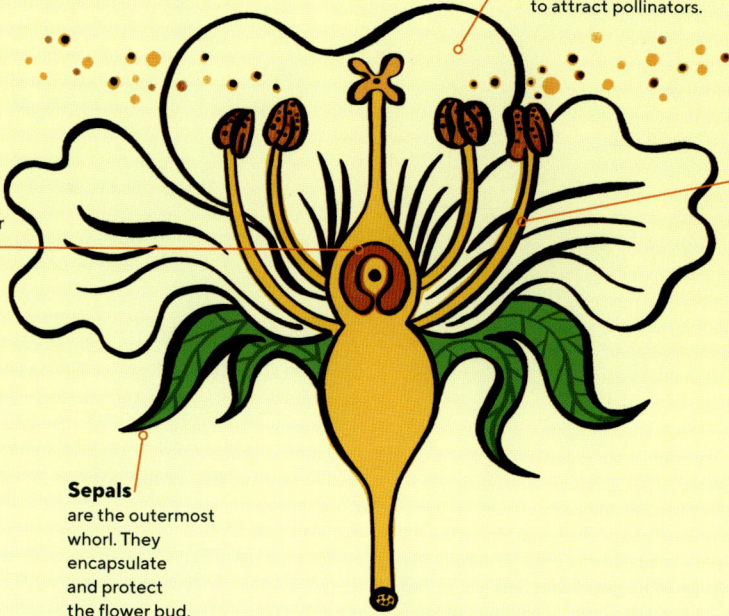

Petals are inner to the sepals. They are usually large, colorful, and showy to attract pollinators.

The most inner whorl of the flower are the **carpels**, which produce the ovules and subsequently the seeds.

Inner to the petals are the **stamens**, which produce pollen.

Sepals are the outermost whorl. They encapsulate and protect the flower bud.

Pollination of Darwin's Orchid

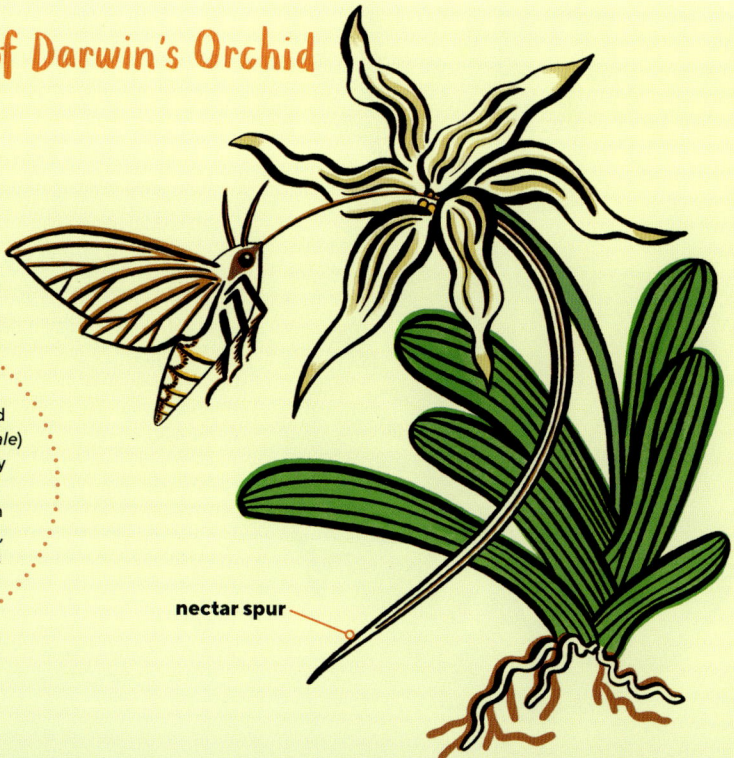

The long nectar spur of Darwin's orchid (*Angraecum sesquipedale*) cannot be accessed by any other species besides a sphinx moth (*Xanthopan morganii*), which has a proboscis nearly a foot long!

nectar spur

10
Can you find a flower on a tree?

When looking for flowers outside, your first impulse may not be to look up. More likely than not, you'll start with short **herbaceous** plants (like tulips, daisies, and lilies; see prompts 92 and 93) or flowering shrubs (like roses, hydrangeas, and forsythia). If you do think of flowering trees, you might imagine showy ones like cherries in the early spring, magnolias, and dogwoods. Or you may think of tropical trees like rosa amarilla (*Cochlospermum vitifolium*) with their spectacular displays of large, colorful flowers that draw animal pollinators like bees. However, the overwhelming majority of tree species around the world are flowering plants. Many are wind-pollinated, though, and don't produce very large or showy individual flowers.

Take a closer look: The smaller tree flowers are like little jewels, hiding in plain sight, awaiting the hand lens of a curious botanist.

TOP: Many familiar trees of the temperate region, like oaks, birches, and this Norway maple (*Acer platanoides*) produce huge numbers of flowers, but they tend to be very small and high in the canopy.

MIDDLE: Black oak (*Quercus velutina*) is wind-pollinated and produces copious amounts of pollen in its catkins.

BOTTOM: Tulip poplars (*Liriodendron tulipifera*) are temperate trees with larger flowers that are a major source of nectar for pollinators.

Cherry blossoms put on a show in spring.

11
Do the first leaves of a seedling look like the mature leaves of the same species?

Maple leaves look like maples, oak leaves look like oaks, and beech leaves look like, well, beeches—except when they don't. The first leaves that emerge from seedlings are called **cotyledons**. In many species, the cotyledons act as the nutritive tissue supporting the development of the young plant. These first leaves come in many shapes and sizes but almost always look very different from the mature leaves.

Cotyledon number follows a relatively strict evolutionary pattern. The grasses and their relatives are called the **monocots**, which means "one cotyledon." All the seedlings of monocots produce a single cotyledon, which emerges when the seed germinates. All other flowering plants have two cotyledons. You may have learned to call them **dicots**, but there are so many diverse plants in this group that it is no longer a preferred term. **Eudicot** is used to described flowering plants with two cotyledons most closely related to the monocots.

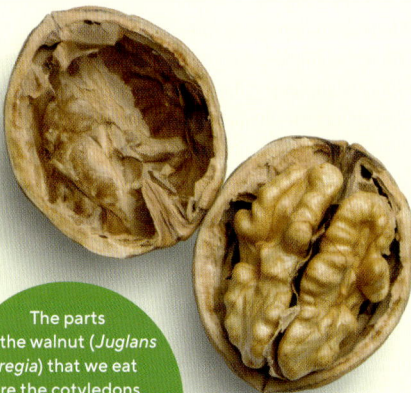

The parts of the walnut (*Juglans regia*) that we eat are the cotyledons. They would have become the first leaves if the seed had germinated.

HAVE YOU EATEN A COTYLEDON TODAY?

Many of our favorite "nuts," like peanuts, walnuts, and almonds, are actually seeds with enlarged cotyledons. Also, the sprouts we put on our sandwiches are generally long stalks with thick cotyledons on the ends. Just as they provide nutrition to the developing seedling, they also provide nourishment for us.

A young beech leaf unfurls in its mature form, looking very different from its cotyledons.

The cotyledons of this American beech (*Fagus grandifolia*) seedling don't look anything like mature beech leaves. Cotyledons are generally thick and fleshy, with a large amount of stored carbohydrates, lipids, or other energy-dense macromolecules.

Southern shorthusk (*Brachyelytrum erectum*) anthers dangle below, while white feathery stigmas await pollen.

12
Can you find the flowers on a grass?

Grasses cover around 30 percent of Earth's land area. But they are not just green carpets to walk on. They are one of the most diverse groups of plants, with nearly 12,000 species in the grass family (Poaceae). They also produce flowers. While grass flowers may not be the showiest you've ever seen, they are actually quite elaborate. However, you will need a hand lens, or even a microscope, to fully appreciate their intricate beauty.

Grasses are highly specialized organisms evolved for pollination by the wind. They have rid themselves of the burden of attracting animals for pollination. They are resilient, strong, and diverse. Remember that the next time you walk on some grass.

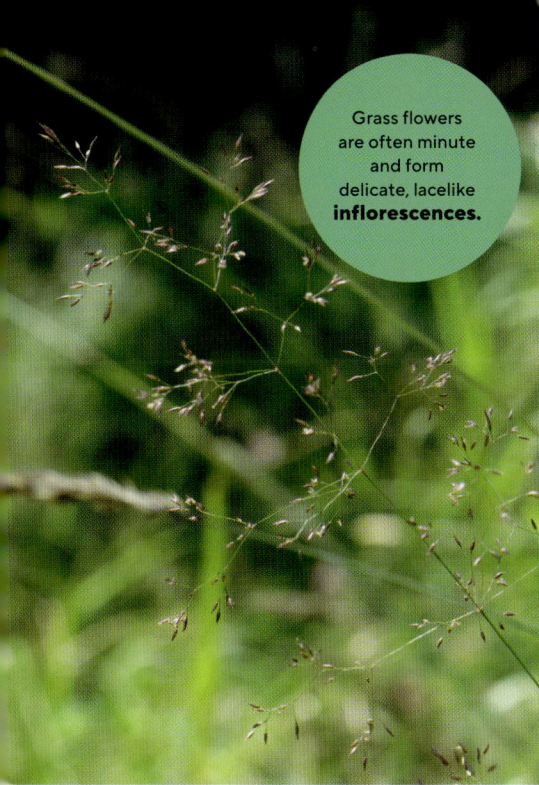

Grass flowers are often minute and form delicate, lacelike **inflorescences.**

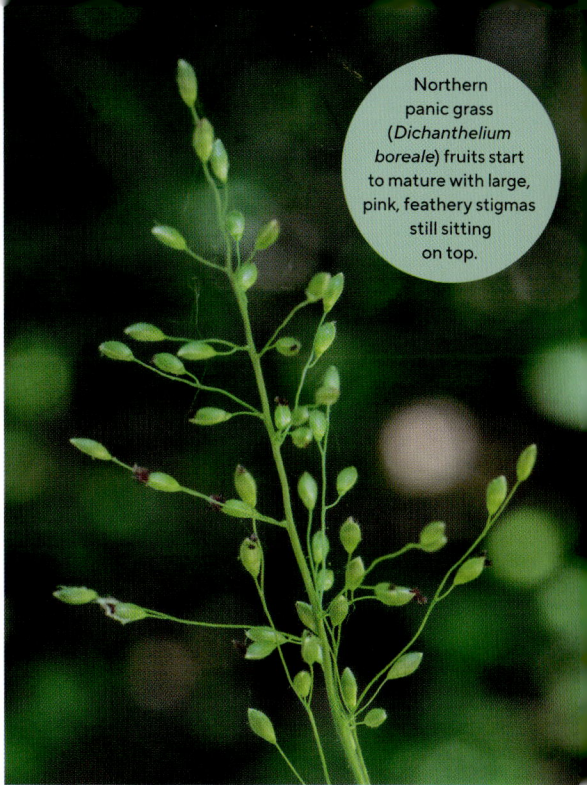

Northern panic grass (*Dichanthelium boreale*) fruits start to mature with large, pink, feathery stigmas still sitting on top.

Parts of a Grass Flower

Each grass flower is composed of highly miniaturized floral parts.

Stigmas are typically grass flowers' showiest parts, large and feathery like kites in the wind—perfect for catching airborne pollen.

Anthers produce copious amounts of pollen for wind dispersal.

The **palea** is the inner papery portion of an individual grass flower surrounding the **lodicules**, which are lumpy, reduced versions of flower sepals and petals.

The **lemma** is the outer papery portion of an individual grass flower.

Glumes are papery, leaflike structures that surround the flower and serve a protective function.

13
What happens when you blow on a fluffy dandelion?

The dandelion (*Taraxacum officinale*) pappus helps the fruits catch the wind (or your breath), allowing them to be carried far distances.

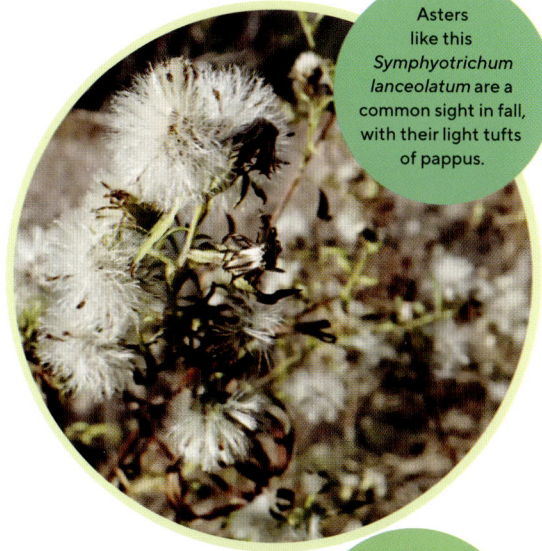

Asters like this *Symphyotrichum lanceolatum* are a common sight in fall, with their light tufts of pappus.

If you grew up in a temperate region, you have surely plucked a dandelion growing in the yard or on a sidewalk and given it a good blow. What happens next captivates young children and botanists alike. Regardless of whether your childhood dandelion wish came true, what surely did happen is that you dispersed the small, seedlike fruits of that individual flower (see prompt 53). But did you ever give much thought to what those structures are?

A dandelion is not a single flower but rather a cluster of dozens of flowers, sometimes called **florets**. After pollination, a dandelion head is full of hard, single-seeded fruits called **achenes**, capped with a white crown of the parachute-like **pappus**. The pappus is a remnant of the **calyx**, the portion of the flower that surrounds the petals.

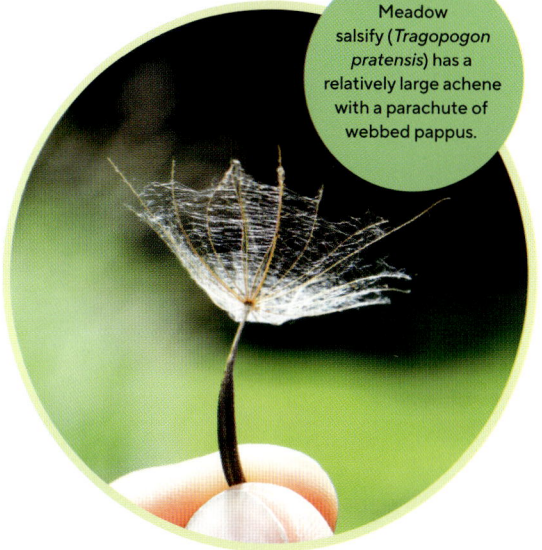

Meadow salsify (*Tragopogon pratensis*) has a relatively large achene with a parachute of webbed pappus.

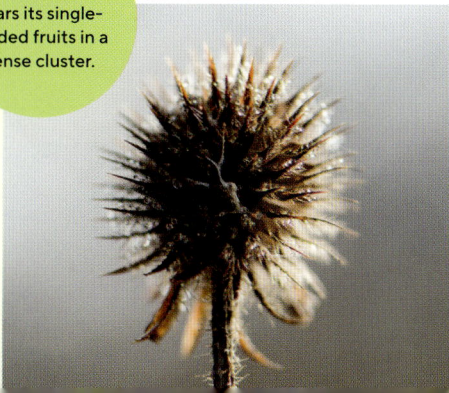

Dipsacus pilosus bears its single-seeded fruits in a dense cluster.

FINDING A WAY

The pappus is not the only modification of the calyx. Across other species, the calyx may be modified into wings for wind dispersal or fleshy structures for dispersal by animals. Anything that helps a plant disperse its seeds will give it a leg up in the polity of nature.

parts

41

14
Can you find a leaf with a smooth edge and a leaf with a serrated edge?

A leaf is made up of a **petiole**, or stalk, and a **lamina,** or blade. The edges of leaves have garnered much attention from plant taxonomists and evolutionary biologists for intriguing reasons. Leaf edges can help us tell apart different species, genera, or families. Oaks often have lobes, and red oaks have sharply tipped lobes. Beeches, dogwoods, and chestnuts have simple leaves with serrated or toothed edges, while magnolias and spicebush have entire margins, meaning they have no ridges, bumps, or teeth. Names have been given to these patterns of leaf margins. For instance, **crenate** describes a bumpy margin, while **dentate** applies to a finely toothed margin, and the list goes on.

Greene's mountain ash
(*Sorbus scopulina*) has serrate margins
(like a serrated knife).

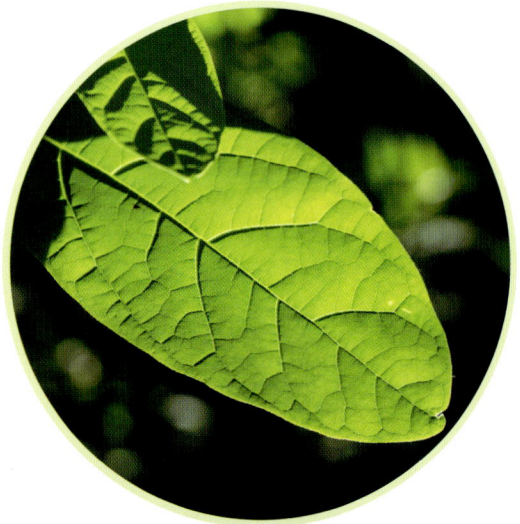

Northern spicebush
(*Lindera benzoin*) has an *entire,* or
smooth, margin.

THE LEAF-EDGE MYSTERY

The leaf-edge pattern that has piqued the interest of evolutionary biologists is that serrated or lobed margins are much more common in temperate lineages compared to tropical ones, which tend to have entire margins. These observations were first made over 100 years ago, and the reason behind this pattern is still unknown. One camp of scientists believes that the edges of leaves help release excess water (**guttation**) in the early temperate spring as leaves are expanding.

On the other hand, some scientists argue that toothed or lobed leaf margins help the leaves fold up more efficiently inside their buds, like origami, allowing for many individual leaves to sit snuggled together in plants with seasonal climates.

parts

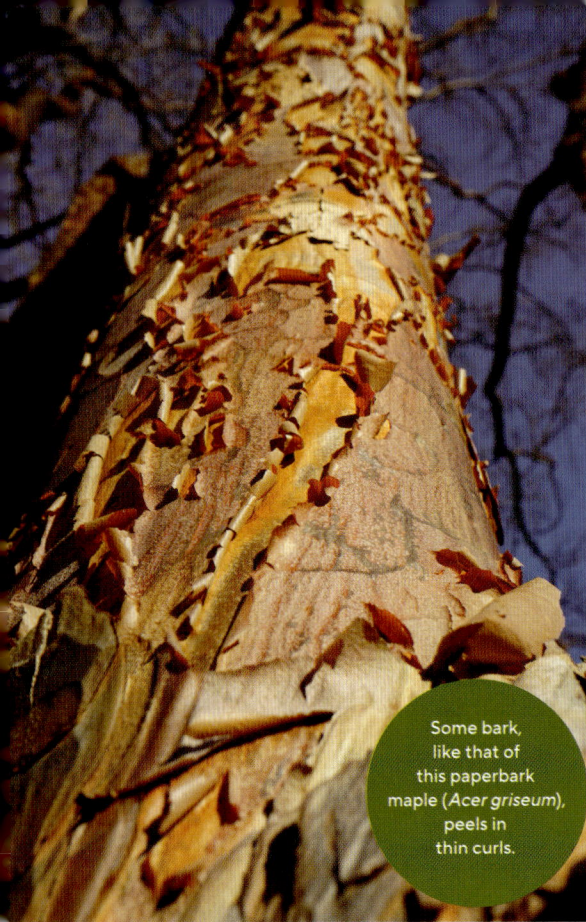

Some bark, like that of this paperbark maple (*Acer griseum*), peels in thin curls.

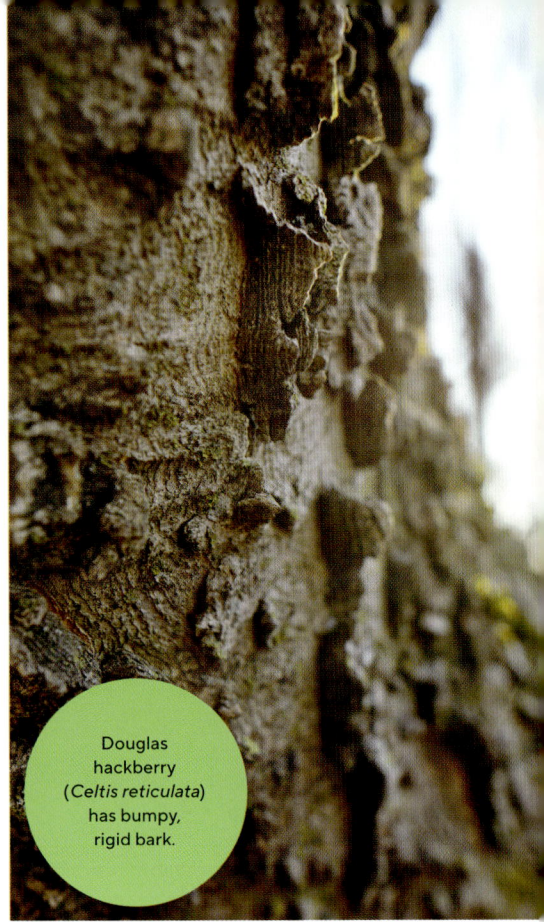

Douglas hackberry (*Celtis reticulata*) has bumpy, rigid bark.

15
How many different textures of bark can you find?

All vascular plants start their lives with an epidermis and a cuticle (see prompt 17) that protect their leaves and stems. However, if the plant is woody, as the individual grows in diameter, the epidermis ruptures. If nothing developed to replace the epidermis as a plant grew, its inner tissues would be exposed to the world. Bark replaces the epidermis over time, becoming the outermost protective layer of a woody plant stem.

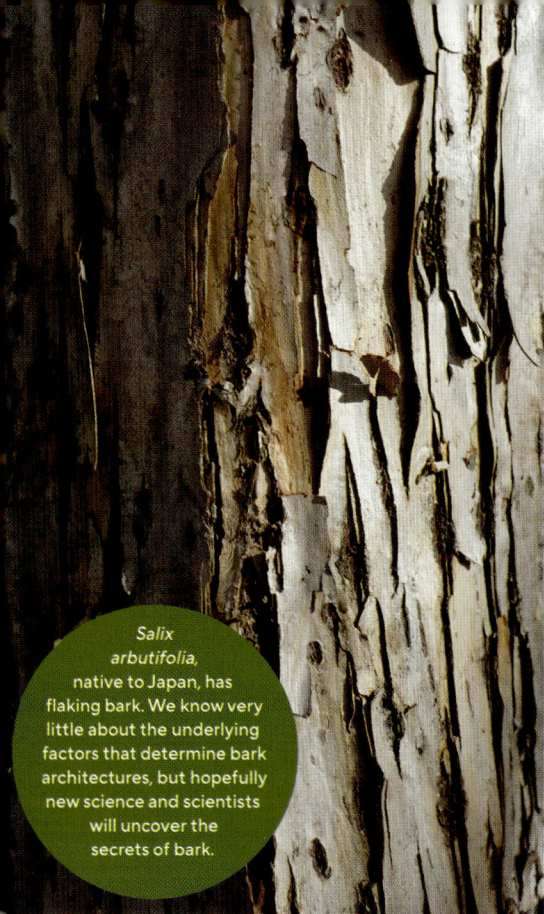

Salix arbutifolia, native to Japan, has flaking bark. We know very little about the underlying factors that determine bark architectures, but hopefully new science and scientists will uncover the secrets of bark.

The bark is a fascinating compound structure made up of multiple sets of tissues derived from different parts of the plant. The outer bark develops from a layer of stem cells called the cork **cambium**, which produces a cylinder around the stem. Rather than a single sheath, this meristem is made up of overlapping plates, somewhat like scales. Bark comes in a diversity of colors and textures.

THE CORK OF CORK CAMBIUM

As you may suspect, cork cambium and cork (the material we make into stoppers for wine bottles) are related. Bark is partially a product of the cork cambium, and corks are made from the especially thick outer bark of a special oak species (*Quercus suber*) that grows in the Mediterranean. The Alentejo region of Portugal is particularly well known for cork cultivation.

16
Can you find the patterns of nodes repeated along a branch?

Each piece of a jigsaw puzzle has a particular shape and fits together with others in a certain way. If you have ever tried to jam two unmatched pieces together, you surely failed. You can use the jigsaw analogy to describe the development of animals. For the most part, they have very particular body plans—an arm here, a leg there, an eye over there. However, plants do not grow like jigsaw puzzles. Instead, they grow like another puzzle-based game called tangrams, where any single piece can fit with any other piece in a multitude of different ways. Even within a single species, the branching architecture of an individual plant will look drastically different from that of its siblings.

The tangram puzzle piece of plant development is called the **phytomer**. Each phytomer is composed of a single **node**, a place where leaves and buds will develop, and **internodes**, or elongated stem regions. These nodes and internodes can be stretched and compressed in many different ways. While the arrangement of branches in space may vary across individuals, the phytomer is the basic repetitive unit across all vascular plants on Earth.

In the Japanese larch, *Larix kaempferi*, some branches produce long internodes, which allow the plant to grow taller and branch out. Other branches produce very compact internodes, leading to extremely short shoots.

The phytomer is the repeating unit of the plant that includes a leaf, node, and internode.

Nodes are the sites of leaf and bud placement. At each node is a resting bud, easy to see on this Japanese elm (*Ulmus davidiana*).

17
What is the waxiest plant part you can find?

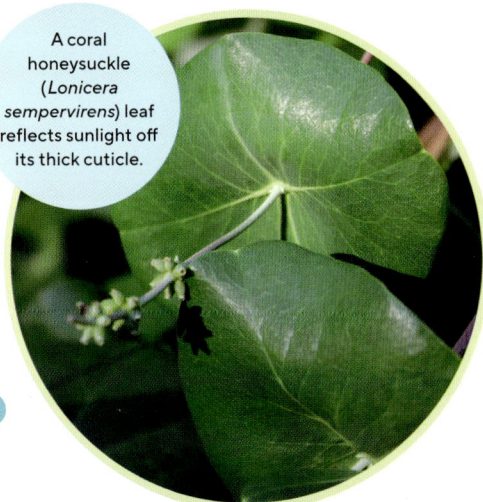

A coral honeysuckle (*Lonicera sempervirens*) leaf reflects sunlight off its thick cuticle.

Plants need water, an obvious fact that becomes tangible to anyone who has returned to their garden or houseplants after a vacation to find them wilted or dead. While some plants can dry out almost entirely and then rehydrate, the majority of land plants must maintain relatively stable hydration. They do this, in part, by covering their entire bodies with a waxy layer called the cuticle. The cuticle is primarily made of a polymer called **cutin**, which does a fantastic job trapping moisture within the plant.

Since leaves are generally the site of photosynthesis and often lose the greatest amount of water given their large surface area, that is where the cuticle is often thickest. However, some species that live in very hot and dry environments have forgone large, transpiring leaves altogether and instead cover their large photosynthetic stems with a thick cuticle. This is often seen in cacti. If you can ever sneak your finger between the sharp spines, you can even wipe the cuticle off. In addition to holding water in, the cuticle acts as a sunscreen, further protecting plants that are exposed to full sun.

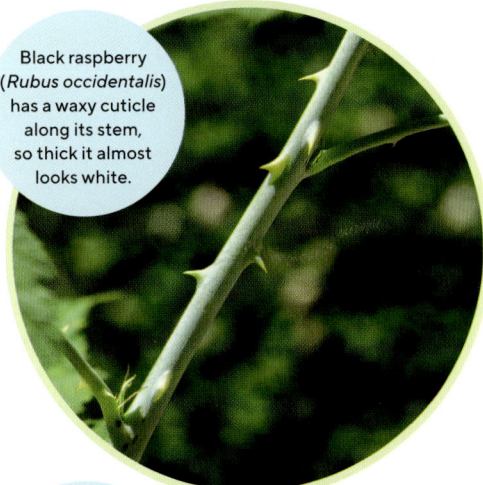

Black raspberry (*Rubus occidentalis*) has a waxy cuticle along its stem, so thick it almost looks white.

Lowbush blueberry (*Vaccinium angustifolium*) has a waxy cuticle that coats its fruit.

Succulent plants often live in sunny, dry environments and can be coated with a thick cuticle to help retain moisture and reflect UV light. This wax can even be wiped off. Can you find the fingerprint on one of these leaves?

JUST ADD HOLES

When plants migrated onto land 475 million years ago, they developed cuticles, possibly to deal with the dry atmosphere. Unfortunately, this also made their bodies impermeable to carbon dioxide, the very molecules they need to survive. No known biological polymer is permeable to carbon dioxide but impermeable to water, because water molecules are smaller than carbon dioxide molecules and have similar properties. To overcome this challenge, plants poked holes in their cuticles by evolving stomata.

parts

Stems of twining plants, like this American hog-peanut (*Amphicarpaea bracteata*), move in slow circles through the air until they touch something solid. Exhibiting **thigmotropism**, or touch-responsive growth, it wraps tightly around its new support.

18
Find two different plants that climb.

How do they grasp onto surfaces?

Living as a vine allows a plant to gain height without investing in building a large, sturdy, free-standing body. Vines exploit the structural soundness of trees, telephone poles, and buildings to support their weight and eke out a living. By racing to climb another plant or structure, vines can escape the shade and gain access to sunlight. This growth strategy works extremely well, and many unrelated groups of plants have independently evolved to grow as climbing vines.

Since many different plants evolved their own way to vine-hood, they have not all learned to climb in the same way. Three of the most common strategies for climbing are twining, **tendrils**, and aerial roots (see prompt 19).

Tendrils are also thigmotropic when they grab hold of a supporting structure. However, instead of wrapping their main stems around the support, these plants produce smaller structures off their main stems that wrap and grab, as is the case with these grapes (genus *Vitis*).

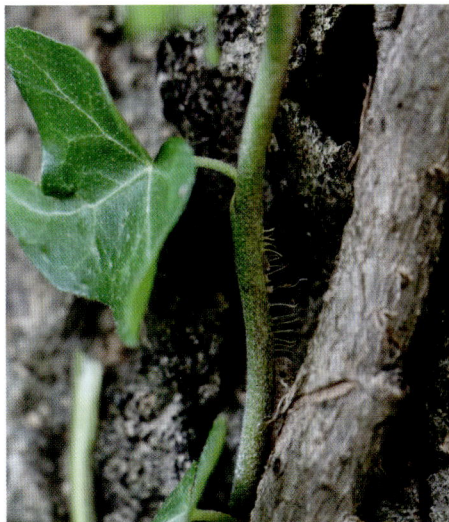

Some plants climb by producing aerial roots along the length of their stems. Many plants that climb with their roots, like English ivy (*Hedera helix*), secrete adhesive substances that can form incredibly strong attachments with their supporting structures. Charles Darwin was actually the first to document this sticky fluid!

parts

19
Can you find a root growing aboveground?

Roots are produced belowground and help plants take up water and nutrients from the soil—at least, that is what most of us have been taught (see prompt 5). But many plants produce roots aboveground on their stems! These are called shoot-borne roots or **adventitious roots**. These roots generally don't serve the traditional function of underground roots but rather facilitate other functions, like climbing or support.

On poison ivy (*Toxicodendron radicans*), small rootlets are produced along the stem and secrete a sticky, cementlike substance. This helps poison ivy creep up trees, fences, lampposts, and old buildings.

Ferns and their relatives grow roots all along their shoot systems. Because this pattern of growth appears across such a broad swath of plants and in some of the oldest vascular plant fossils, we can infer that the common ancestor of all vascular plants likely produced roots and shoots from the same system.

Strangler figs (*Ficus*) germinate high in the canopies of other trees and develop roots that grow to surround the host tree's trunk over time. Eventually, the host tree will be smothered and die, leaving behind a hollow new trunk made up of thick, intertwined fig roots.

20
How far from the trunk can you follow a tree root?

Swamp birch (*Betula alleghaniensis*) and eastern hemlock (*Tsuga canadensis*) send their long roots over a boulder in search of soil and water.

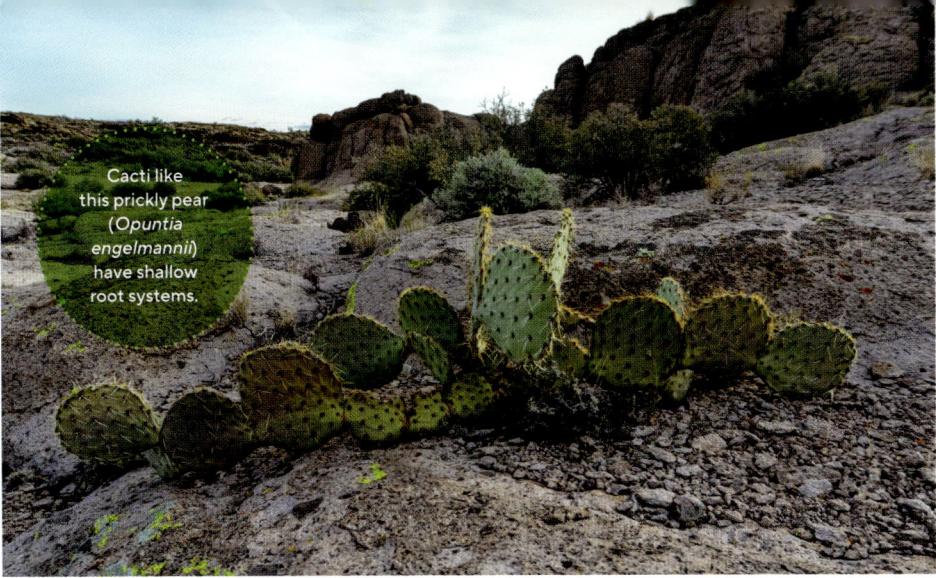

Cacti like this prickly pear (*Opuntia engelmannii*) have shallow root systems.

The shoot system grows in predictable patterns. But when we dive belowground into the root system (see prompt 19), things are a bit different. As roots search for water and mineral nutrients, they grow outward into lateral roots. Lateral roots often grow somewhat haphazardly from the main root because the soil (or other substrate) is not as uniform as the air aboveground. One part of the soil may be wet, another dry, another has a rock in the way, one spot has high nitrogen right next to a spot with low nitrogen. Nonetheless, there is still some predictability in root architecture across species and in different habitats.

The depth of the water table is one of the most important ecological factors affecting rooting. Roots often won't penetrate deep into groundwater because they can't survive the lack of oxygen in the saturated soil. Roots must do their job of stabilizing the plant and obtaining water and nutrients, all while enduring a harsh underground world. Given the difficulty of studying roots, there are many secrets of the belowground root system that we have yet to discover.

Root Depth

Little bluestem grass (*Schizachyrium scoparium*) and many other grasses often have deep and fibrous root systems to penetrate the soil. In some cases, these grass roots can be several feet deep. Cacti have shallow roots that stay close to the surface to suck up water as soon as it rains.

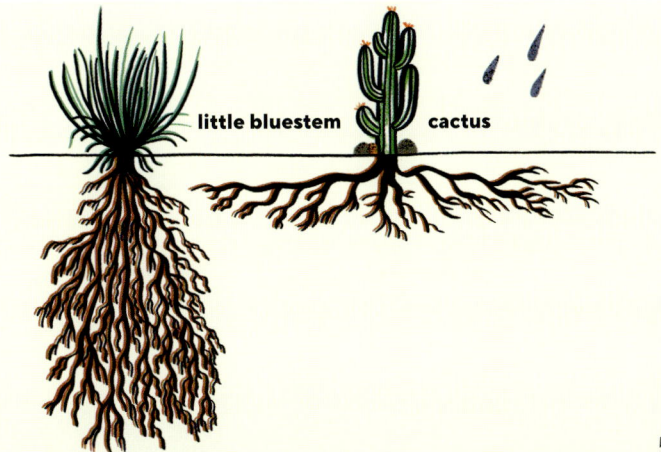

little bluestem   cactus

parts

21
Compare the textures of several different leaves.

Are they smooth, soft, or hairy?

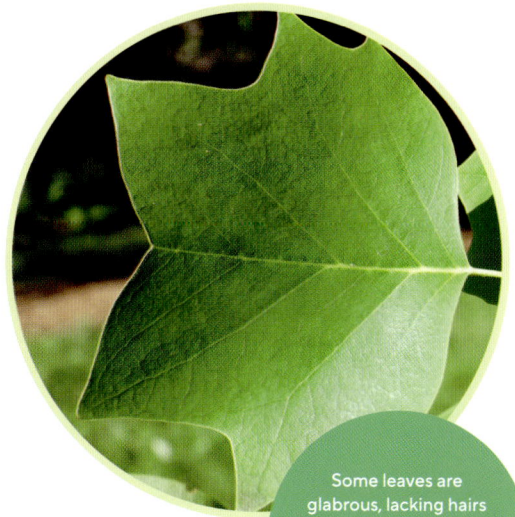

Some leaves are glabrous, lacking hairs or other projections, like this tulip poplar (*Liriodendron tulipifera*) leaf with a waxy cuticle.

Leaves are like snowflakes or fingerprints—no two are exactly alike. Of course, one leaf will be more like other leaves on the same individual plant compared to a different one—and this holds true within populations and species (but see prompt 54). Between species, however, this variation can be immense. Run your finger across the leaf surfaces of several species and you'll notice a remarkable difference in texture. When you feel the surface of a leaf, it may be smooth, soft, or prickly. Watch out—some even have sharp, needlelike hairs called **trichomes** that may inject you with irritating histamine (think stinging nettle).

The leaf is the powerhouse of the plant body, and its form relates directly to its function. A waxy cuticle helps retain water. Pubescent (fuzzy) leaves can help modulate temperatures or sunlight. And sharp trichomes can serve a defensive role. There are many "just-so stories" about leaf textures. Some botanists may argue that hairy leaves protect from this or that. However, it may not be the case that every leaf modification is fine-tuned by natural selection. Rather, the plant may have inherited traits that neither help nor harm it, allowing it to survive just fine the way it is.

Some trichomes can be glandular, containing numerous chemical compounds secreted from their tips, making the leaf sticky. This *Cannabis sativa* leaf has gland-tipped trichomes that secrete terpenes and psychoactive chemicals.

A tomato (*Solanum lycopersicum*) leaf has minute, gland-tipped trichomes and irritating, needlelike hairs.

A giant salvinia (*Salvinia molesta*) has eggbeater–shaped trichomes that repel water and keep the leaves of this aquatic plant afloat.

Not all **evergreen** plants have needles. Some, like this *Rhododendron maximum*, have broad leaves that are often thicker or more robust to survive the winter.

22

Examine the broad leaves or needles of an evergreen tree.

What characteristics might help them survive the winter?

The needles of the eastern hemlock (*Tsuga canadensis*) are narrow with thick cuticles. This, combined with a low density of stomata, helps limit water loss.

Leaves of most species are deciduous—they are produced in a single growing season and last until the first frost or hard drought of the year. If you go out botanizing in the winter or the dry season (depending on your climate), the only green leaves are evergreen plants—those that hold on to their leaves year-round.

Developing leaves that last a single growing season is an energetically low investment compared to developing those that last several years. **Deciduous** leaves are cheap. Evergreen leaves—whether needles like those of an eastern hemlock (*Tsuga canadensis*) or broad leaves like those of a great laurel (*Rhododendron maximum*)—are built differently. They are often thicker, have a waxier cuticle, and are structurally more robust. These leaves must endure the elements year after year, including harsh temperate winters or hot dry seasons. The cold is not the only issue in winter—water in the soil can be locked up as ice, making it difficult for plants to stay hydrated. This means that evergreen leaves must be able to survive drought conditions in addition to the cold temperatures.

parts

59

23

Can you find a leaf with parallel veins?

HINT: Look for grasses.

Many plants have vascular systems much like humans.

The difference is that we use muscles in our hearts to pump blood throughout our bodies. This takes energy and means that the blood in our veins is under positive pressure (it will spew out if we are injured). In contrast, plants move water through their veins passively, without any energy input. As water molecules evaporate from the stomata, other water molecules are pulled through the veins.

This process relies on the size and patterning of a plant's veins. With more veins, a plant can move more water and increase its photosynthetic rate. Leaf-vein architecture is quite varied across the plant tree of life (see prompt 1).

Many flowering plants have complex, reticulate veins that move water throughout the leaf in crisscrossing networks. Branching veins also provide structural support, allowing for broad leaves.

False Solomon's seal (*Maianthemum racemosum*) produces relatively broad leaves for a monocot, which remain sturdy thanks to arcing bundles of parallel veins.

Grasses and their relatives, which are monocots (see prompt 11), typically have simple, linear veins that run parallel to each other in long, narrow leaves. Water is also able to move across these leaves, thanks to leaky vascular tissues.

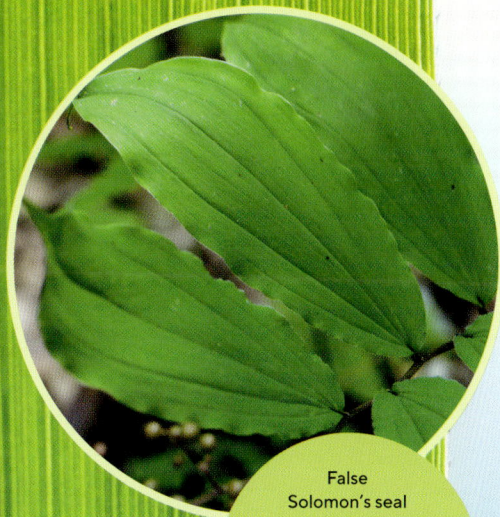

24
Can you see the veins in the petal of a flower?

Hold a flower petal up to the light and you should be able to make out the telltale meanderings of veins. Just like leaves, stems, and roots, flower petals are made of living tissue that needs water, sugar, and other nutrients to grow. These supplies are delivered by the vascular system—veins that include xylem and phloem tissues.

The fact that both flower petals and leaves are generally flattened structures crisscrossed by veins is not just a coincidence. They are similar because flower petals are evolutionarily modified leaves. Large, flat leaves evolved to harvest sunlight for photosynthesis, but over time flowering plants co-opted leaves to make up the various parts of their novel reproductive apparatus, the flower.

Petals and sepals, the outermost structures of flowers (see prompt 9), are typically the most leaflike. They are often broad and flat, and many sepals are green and able to perform photosynthesis; they even have stomata. As flowers evolved, their petals became specialized to attract animal pollinators. Now most petals produce little to no green chlorophyll but myriad other pigments, yielding colors from red to purple and everything in between.

Veins in the petals of Virginia spring beauty (*Claytonia virginica*) are highlighted in purple.

Furrowed trillium (*Trillium sulcatum*) has veins embossed in the petals.

The leafy origins of sepals and petals may be relatively easy to imagine, but the innermost structures of a flower, the pollen-bearing stamens and ovule-bearing carpels, are also evolutionarily modified leaves.

Korean mountain magnolia (*Magnolia sieboldii*) has a large single flower.

25
Can you find one plant with singular flowers and one with clusters of flowers?

Even if you see yourself as a fern person, pine person, moss person, or mushroom person, you may have a special flower close to your heart. Indeed, flowers are the muses of poems, songs, and fairy tales. Roses have been adopted to express love, and tulips have been recruited for Easter. But the way a plant packages its flowers is quite variable.

Some species bear flowers in single units, like the roses or tulips, while others are born in clusters that we call inflorescences. These inflorescences can be loosely arranged (sometimes called "relaxed"), like the common snapdragon (*Antirrhinum majus*) or butter-and-eggs (*Linaria vulgaris*), where each individual flower can be easily identified with the naked eye. In other cases, large inflorescences can act as a single flower, like in the sunflower (*Helianthus annuus*) or *Viburnum*.

Cyathium Inflorescence

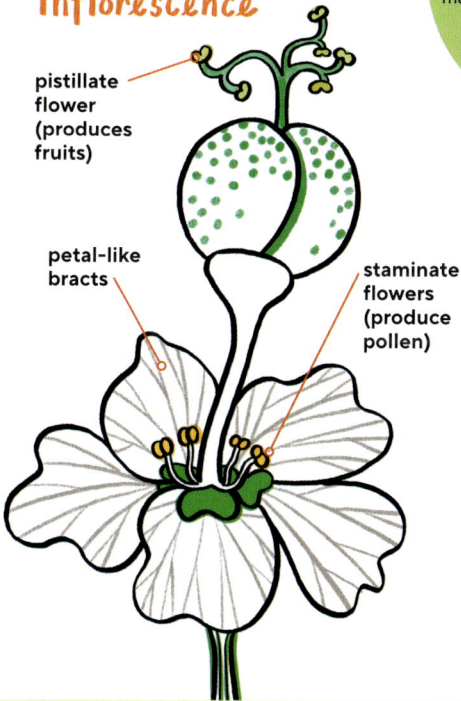

pistillate flower (produces fruits)

petal-like bracts

staminate flowers (produce pollen)

Plants in the spurge family, Euphorbiaceae, have figured out how to create a single flower made up of individual flowers. A cyathium inflorescence has a cluster of single pollen-producing flowers and a single ovule-producing flower. With a few other auxiliary parts, this inflorescence acts as a single flower.

Red clover (*Trifolium pratense*) has individual purple flowers clustered together into a compact, globe-shaped inflorescence.

In this Jerusalem artichoke (*Helianthus tuberosus*) and other flowers in the sunflower family (Asteraceae), hundreds of individual flowers in the center make up one large inflorescence that mimics a single flower and makes a highly visible target for pollinators.

A field garlic (*Allium vineale*) inflorescence opens its clustered flowers.

parts

26
Examine how leaves are arranged around a stem.

Are the leaves directly opposite each other, or do they have some other pattern?

The spatial arrangement of leaves along a stem is called **phyllotaxy**. All leaf arrangements fall into one of three general categories: alternate, opposite, or whorled.

If a leaf stands alone, with no other leaves emanating from that point on the stem, that plant has alternate leaves. If leaves occur in pairs along a stem, that plant has opposite leaves. An opposite leaf always has one partner 180 degrees around the other side of the stem. If more than two leaves grow around a single point on a stem, that plant has whorled leaves.

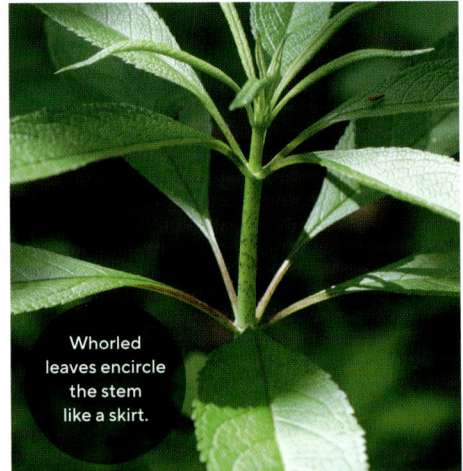

Whorled leaves encircle the stem like a skirt.

USEFUL EXCEPTIONS

It is unclear whether these patterns of phyllotaxy ever confer an evolutionary advantage, but they are usually consistent at the species, genus, and even family level and can be useful for plant identification. In fact, some species are named based on their exception to the rules of their group. We're looking at you, alternate-leaved dogwood (*Cornus alternifolia*).

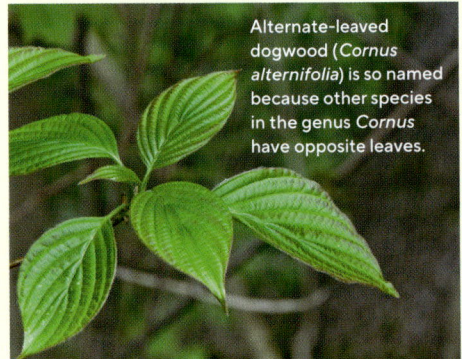

Alternate-leaved dogwood (*Cornus alternifolia*) is so named because other species in the genus *Cornus* have opposite leaves.

Alternately arranged leaves zigzag their way down the stem, like on this smooth Solomon's seal (*Polygonatum biflorum*). In this case, because the leaves are held in a single plane, they are also called **distichous.**

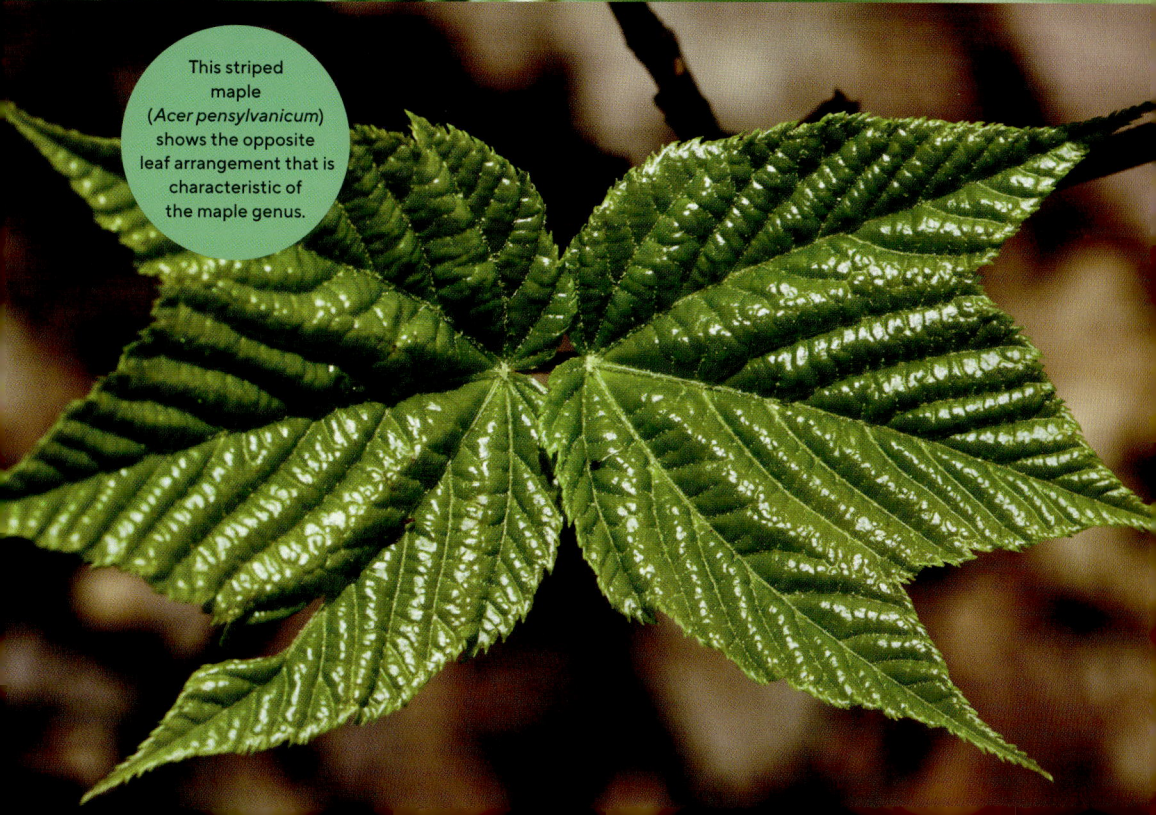

This striped maple (*Acer pensylvanicum*) shows the opposite leaf arrangement that is characteristic of the maple genus.

27

Can you find plants with three different kinds of sharp projections?

Spines are evolutionarily modified leaves. Cacti (family Cactaceae) have spines spaced regularly across their swollen, green stems, as in this desert prickly pear (*Opuntia phaeacantha*).

Prickles are simply sharp outgrowths of the outermost layer of tissue on the stem, the epidermis—imagine if your hair were prickly! Prickles tend to be distributed more randomly across the plant body—a hint to their developmental origin. A sandbox tree (*Hura crepitans*) has numerous prickles growing out of its bark.

Animals must eat plants (or eat other animals and fungi that eat plants). In order for plants to survive, grow, and reproduce, they must photosynthesize and limit the damage done by herbivores. Plants are stationary organisms, and one way they have evolved to protect themselves is by growing sharp projections from their body. In fact, this strategy is so effective that it has independently evolved many times across the plant tree of life. Not all plants develop these sharp projections in the same way. Plants can have **spines**, **thorns**, or **prickles**, which are all sharp objects that develop from a different plant organ.

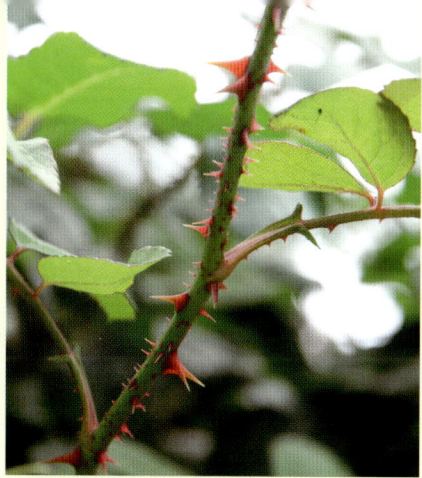

A THORN BY ANY OTHER NAME . . .

Roses technically do not have thorns. The sharp structures that pierce your finger are actually prickles. But somehow that's not quite as poetic.

Thorns are modified branches that grow out of a main stem. They are often found at nodes, just above leaves (see prompt 16). They can even be branched themselves, as is often the case on honey locust trees (*Gleditsia triacanthos*).

28

Look for outgrowths where a young leaf attaches to a stem.

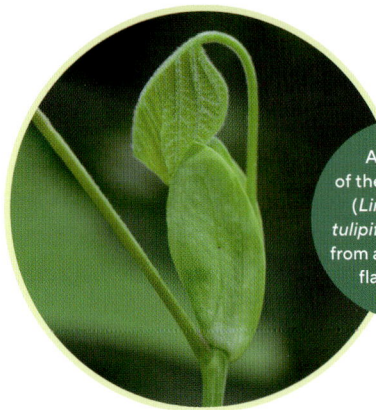

A new leaf of the tulip poplar (*Liriodendron tulipifera*) emerges from a pair of large, flat stipules.

HINT: Focus your attention on the newest growth toward the tips of branches.

Trace a young leaf from tip to base. Where the leaf attaches to the stem, you may see a small, leafy outgrowth. This is called a **stipule**. Stipules are not present in all plants. When they do appear, they may be ephemeral, only present while a leaf is expanding. In fig species (*Ficus*), the stipule completely surrounds the developing leaf and has the appearance of a dunce cap. Likewise, in the tulip poplar (*Liriodendron tulipifera*), a pair of stipules envelopes the developing leaf as it expands. In these cases, stipules likely help protect the young developing leaves.

Stipules have taken on many forms across the flowering plants and are quite helpful for identifying large plant groups. In some cases the stipules have clear functions, while in other cases they may just be remnants of past developmental programs. Not much is known about the function of stipules, making them another mystery of the botanical world.

Stipules are not always leafy. The black locust (*Robinia pseudoacacia*), shown here, has stipules modified into woody spines that stab any would-be herbivore. Its species name pays homage to the famous spines of acacias.

In the greenbriar (*Smilax*), stipules are modified into tendrils for climbing.

Heart-leaved willow (*Salix eriocephala*) has large, toothed, kidney-shaped stipules at the base of each leaf.

29
Can you find a scar on a branch where a leaf used to be attached?

Horse chestnut (*Aesculus hippocastaneum*) has distinctive horseshoe-shaped leaf scars.

A shagbark hickory (*Carya ovata*) has a large, heart-shaped leaf scar and obvious dots where leaf veins were connected to the stem.

Several leaf scars mark the end of a branch in a hybrid magnolia (*Magnolia acuminata × denudata*).

You can usually find the exact location where a fallen leaf used to be attached to a stem by looking for the mark, or **leaf scar**, that is left behind. Most long-lived plants, even conifers and other evergreens, will eventually drop old leaves and replace them with new ones. When leaves drop on their own (when they are not torn off the plant by wind or animals), they often fall off in a highly predictable way that leaves a clean and sealed-off footprint where the petiole was connected. The process takes place in a layer of cells at the base of the petiole called the **abscission zone**.

Within a species, leaf scars usually have very consistent shapes and patterns of dots that indicate where vascular bundles of xylem and phloem once connected the leaf to the rest of the plant. This makes leaf scars very helpful in identifying plants in the winter or dry season when leaves and flowers may not be around.

The Abscission Zone

The process of abscission happens in several phases. First, a plant will typically break down any useful chemical compounds that can be recycled from the leaf and absorb them to be used elsewhere. Then, a layer just below the abscission zone develops into a corky, protective layer that is reinforced with wax and tough plant polymers. Finally, the cells just above this protective layer die and break away, allowing the leaf to fall off. A waterproof seal forms at the newly exposed surface—the plant does not want an open wound.

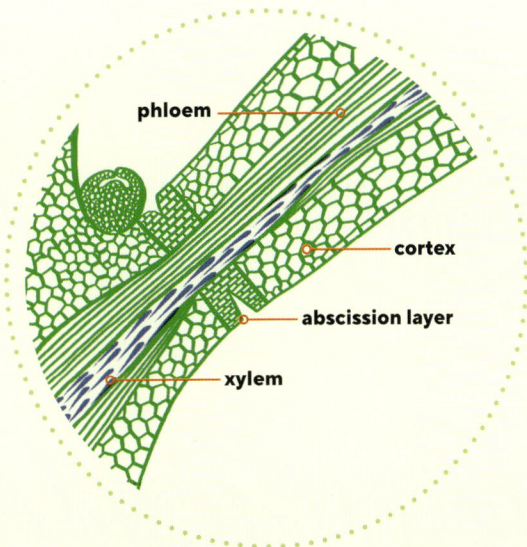

phloem

cortex

abscission layer

xylem

30
Explore the detail of a moss leaf.

Mosses are one of the three clades of bryophytes—the closest living relatives to the rest of land plants. The other two **clades** are the hornworts and the liverworts. Mosses produce leaves, but they are not quite the leaves we are used to seeing. For one thing, they are often microscopic, lacking the intricate layers of most vascular plant leaves (see prompt 4). Also, each moss leaf is no more than a few cells thick.

While these leaves can be appreciated with a hand lens, a dissecting microscope offers a more detailed look at their hidden structures. Because they do not produce true vascular tissues like the ferns, lycophytes, or seed plants, the **bryophytes** are often diminutive in comparison. Not worse, not less adapted, not more primitive—simply smaller. We have to work harder with our poor human eyes to appreciate their beautiful and minute structures.

Some mosses have elaborate leaves, like the serrated leaflets of saber tooth moss (*Plagiomnium ciliare*).

The leaves of sphagnum moss (*Sphagnum cristatum*) have specialized cells to hold water.

Ptychostomum species have translucent, thin leaves.

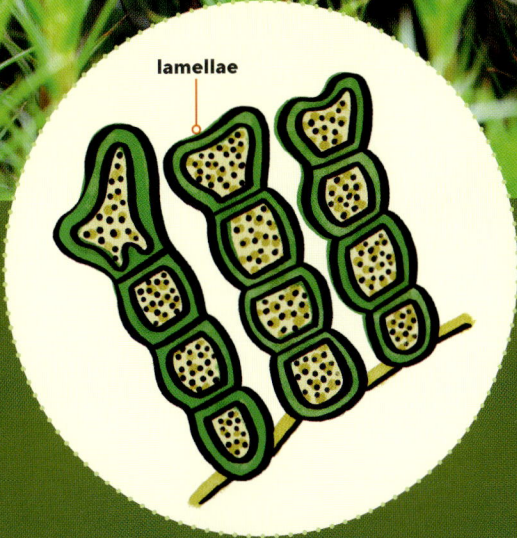

Lamellae

In many moss species, like the common haircap moss (*Polytrichum commune*), the tops of the leaves are lined with columns of several cells; these structures are called **lamellae**. The lamellae may increase the surface area and help facilitate gas exchange.

lamellae

31

Can you find where a moss produces its spores?

Plants originated in the water. Algae, the common ancestors of all land plants, produced **zygotes** (a fusion of sperm and egg) through fertilization directly on their plant bodies. These zygotes would be dispersed and eventually undergo **meiosis**, a word that may bring back memories from biology class. Meiosis is the process of cell division that leads to four new cells, each of which has half of the genetic material of the parent cell. This is how humans produce sperm and eggs. Plants do it, too.

Once plants jumped onto land 475 million years ago, something spectacular happened. Instead of dispersing zygotes, plants retained them on their bodies, providing nourishment while the zygote developed. This allowed the zygote to become an **embryo** (a multicellular product of fertilization), which is where land plants get their formal name: **embryophytes**.

In land plants, embryos undergo meiosis to produce large multicellular structures called **sporophytes**, where the spores are produced. In some of the earliest land plants, and the modern bryophytes, these sporophytes are directly attached to the mother plant and receive nutrients and water through a structure called a foot—similar to an umbilical cord.

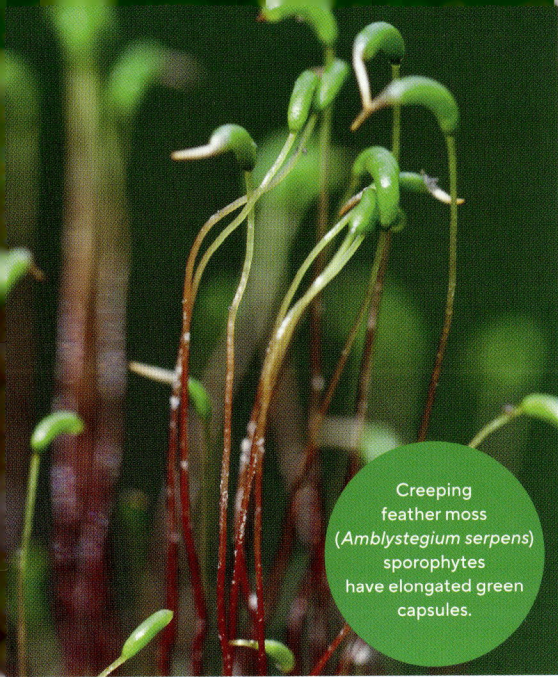

Creeping feather moss (*Amblystegium serpens*) sporophytes have elongated green capsules.

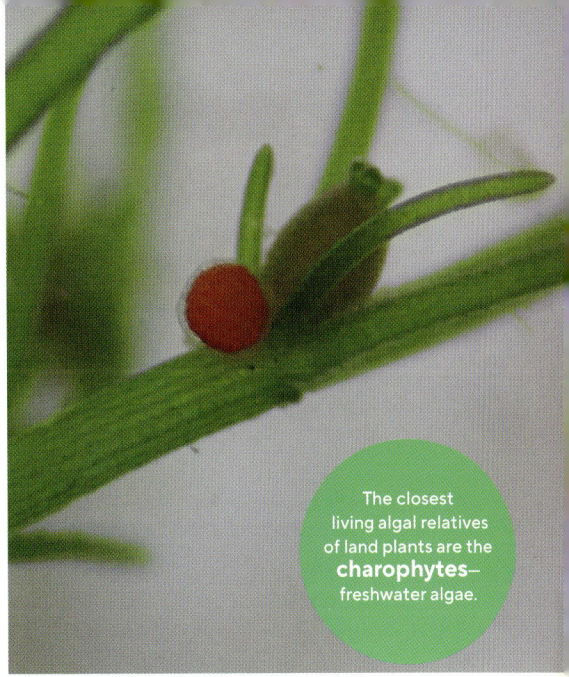

The closest living algal relatives of land plants are the **charophytes**—freshwater algae.

Spore Dispersal

Inside the capsule of the sporophyte, the spore mother cell undergoes meiosis to produce many spores that will, hopefully, be dispersed by wind, land in moist rocky crevices, and germinate into new moss plants.

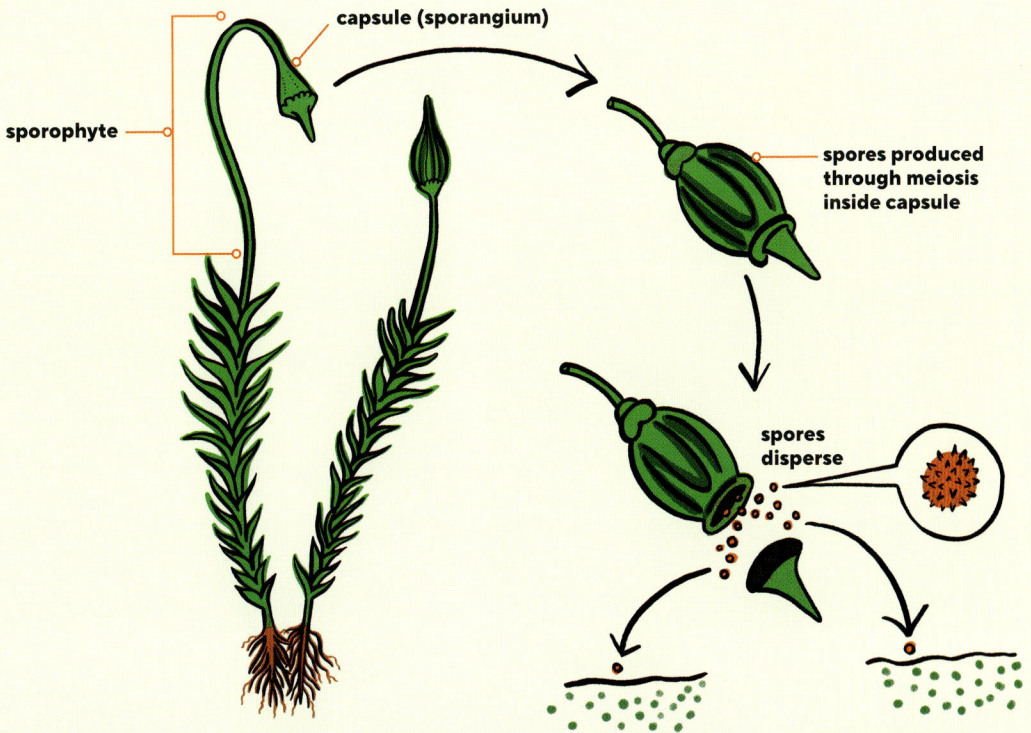

capsule (sporangium)

sporophyte

spores produced through meiosis inside capsule

spores disperse

32
Can you find where a fern produces its spores?

If you have ever looked at the underside of a fern leaf, you may have been mesmerized by the hundreds of tiny dots or zigzag lines smeared across the surface. These are not bug eggs or dirt, but thousands of tiny **sporangia**, the structures that produce spores. Sporangia in nearly all living ferns are produced on the undersides of leaves (sometimes off to the side). While there are exceptions, rarely do we have such a strong and consistent pattern in biology—the scientific field of exceptions.

When sporangia are clustered together into groups, they are called **sori** (singular **sorus**). Sori come in many shapes and arrangements and can be helpful for identifying particular families, genera, and species.

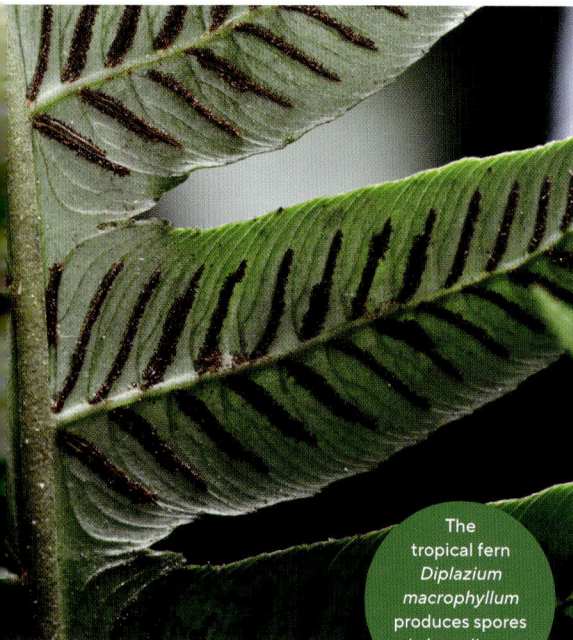

The tropical fern *Diplazium macrophyllum* produces spores in long, linear sori.

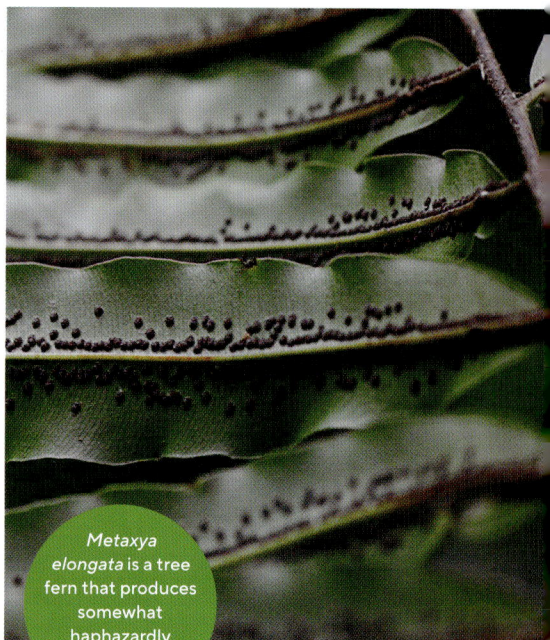

Metaxya elongata is a tree fern that produces somewhat haphazardly spaced sori.

Interrupted fern (*Claytosmunda claytoniana*) bears spores in a set of highly modified leaflets (sporophylls) that look like they are covered in dark beads.

33

Can you find a flower without apparent sepals?

Some flowers, like many in the poppy family (including the greater celandine, *Chelidonium majus*), make **caducous** sepals, which fall off during the blooming process, so the mature flowers appear to lack them.

Sepals are the outermost whorl of a typical flower, found just outside of the petals (see prompt 9). They are usually green and leaflike and enclose the developing flower bud, opening to reveal the mature flower when it blooms. However, not all flowers bear sepals that follow this pattern.

Dutchman's pipevine (*Aristolochia macrophylla*) produces flowers without apparent sepals because the tubular floral structure is composed of fused sepals (not petals). The fact that no whorl of organs is found outside of the pipelike flower provides the major clue that this flower is actually made up of sepals.

Lilies and tulips produce sepals that are as colorful and attractive as their petals. In species like these, where only their relative position (inside or outside) distinguishes sepals from petals, we use one word—**tepals**—to refer to the sepals and petals together.

34
Track a flower until it fruits.

Does it produce one or many fruits? You may be able to see flowers and fruits at various stages of development on a single plant.

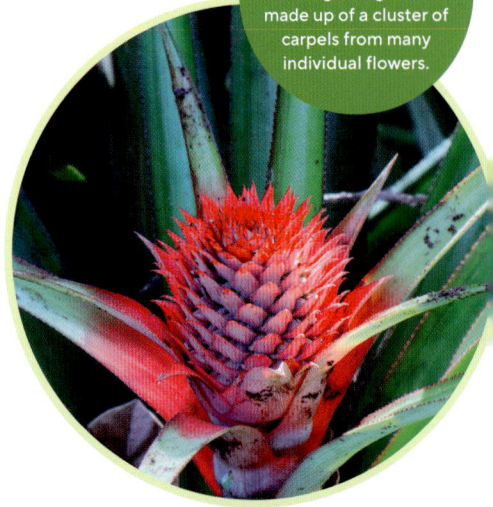

A multiple fruit, like a pineapple (*Ananas comosus*), is a large, single fruit made up of a cluster of carpels from many individual flowers.

Clusia obdeltifolia, shown here, produces spherical fruits that each contain five fused carpels, which are visible when you cut them open. It also drips latex, a feature of the family Clusiaceae.

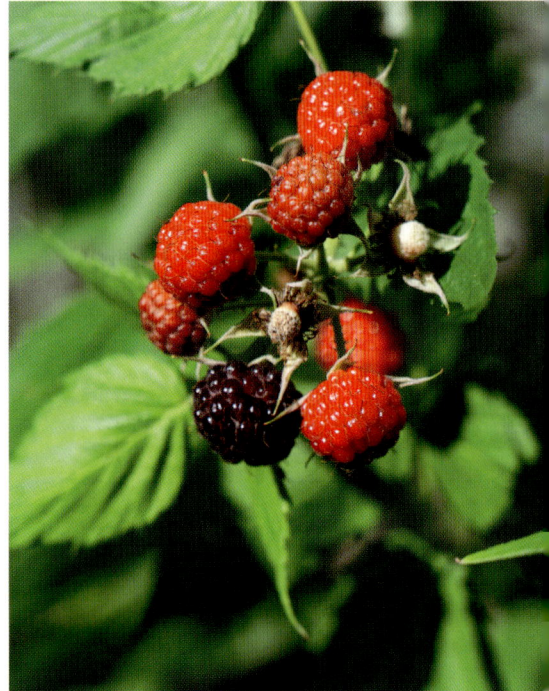

Black raspberry (*Rubus occidentalis*) produces aggregate fruits consisting of many loosely fused carpels all derived from a single flower.

parts

A fruit is part of a flower, the part that protects the developing seeds. What we often think of as a culinary "fruit" represents a tiny fraction of all fruit diversity on our planet. Indeed, fruits come in many textures, shapes, sizes, and colors. Some are soft and fleshy like our delicious North American blueberries, while others are hard and woody, like acorns (see prompt 53). Others, like jewelweed (*Impatiens capensis*), burst open at the slightest touch!

Many traits determine the final form of a mature fruit, but two important traits are the number and fusion of the carpels. Some fruits are large, single organs with multiple fused carpels—think oranges or tomatoes (each space is an individual carpel). Other fruits produce separate simple carpels from a single flower. These include pawpaws (*Asimina triloba*) and buttercups (*Ranunculus*). Still others are **aggregate fruits** with loosely fused carpels or **multiple fruits** made from multiple flowers (think pineapples).

The Fraser magnolia (*Magnolia fraseri*) produces large, brilliantly colored aggregate fruits consisting of many fused carpels from a single flower.

Jewelweed, also known as touch-me-not (*Impatiens capensis*), has explosive fruits that suddenly split open when touched.

35
Describe the color of a newly emerging leaf.

In a region marked with **seasons,** there is nothing like the excitement of newly emerging leaves to signal the beginning of the growing season. In many perennial plants, these leaves have been lying dormant, waiting for the perfect conditions. Most of these leaves, in fact, were made during the previous summer and packed away tightly in resting buds. At the start of the growing season, plants pump water into the cells of these microscopic leaves, expanding them rapidly (see prompt 61).

These brand-new leaves are usually much lighter shades of green than the fully expanded leaf will be. This is because chloroplasts in the leaf (the chlorophyll-containing, light-harvesting organelles that drive photosynthesis) are not fully developed yet. As leaves expand and grow, they produce the deep-green chlorophyll pigments that darken them.

Pale yellowish eastern redbud (*Cercis canadensis*) leaves expand in the spring from a dormant bud. They will darken as chlorophyll develops.

Light green American beech (*Fagus grandifolia*) leaves unfold. Most of the expansion of a leaf is due to water rushing into the cells, not cell division.

Some young leaves will even be red, like those of this black oak (*Quercus velutina*). These shades of red come from different pigment molecules called **anthocyanins**.

Norway spruce (*Picea abies*) has young cones that are as bright pink as a flower.

36
Can you find a young pine cone?

HINT: It won't be brown and brittle.

Pine cones don't always start off as hard, woody structures like the ones hanging over the mantle. Similar to other plant organs, their humble beginnings are as small clusters of cells differentiating in the shoot apical meristem (see prompt 3). Once development begins, a pine cone that emerges in the spring will be soft and fleshy. It may even be vibrantly colored with reds, yellows, pinks, and purples. These colors are often generated by the same pigments (anthocyanins and **carotenoids**) that color the petals of flowers and the flesh of fruits.

Unlike in flowers, these colors are likely not for attracting pollinators, since nearly all conifers are wind pollinated. Rather, the pigments may help shield vulnerable young pine cones from excess UV light, sort of like sunscreen (see prompt 63).

Throughout their development, pine cones open and close for different reasons. When they are young, the scales open to receive pollen. As each ovule is pollinated, the cone scale closes to protect developing seeds. Unlike the development of many flowering plants, seed and cone development in conifers can take several years from pollination. Once a cone is mature, the scales open in response to environmental cues like humidity, or even fire, to release the cone's seeds.

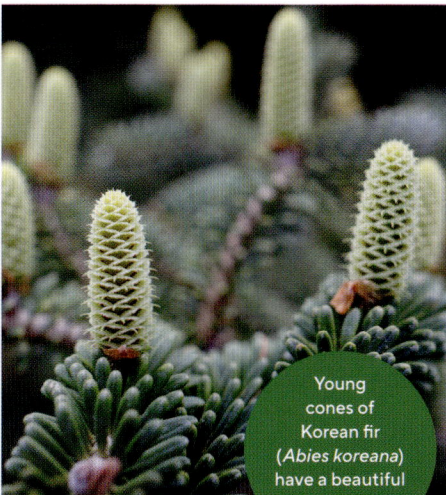

Young cones of Korean fir (*Abies koreana*) have a beautiful golden hue.

Young cones of Dahurian larch (*Larix gmelinii*) show contrasting patches of color.

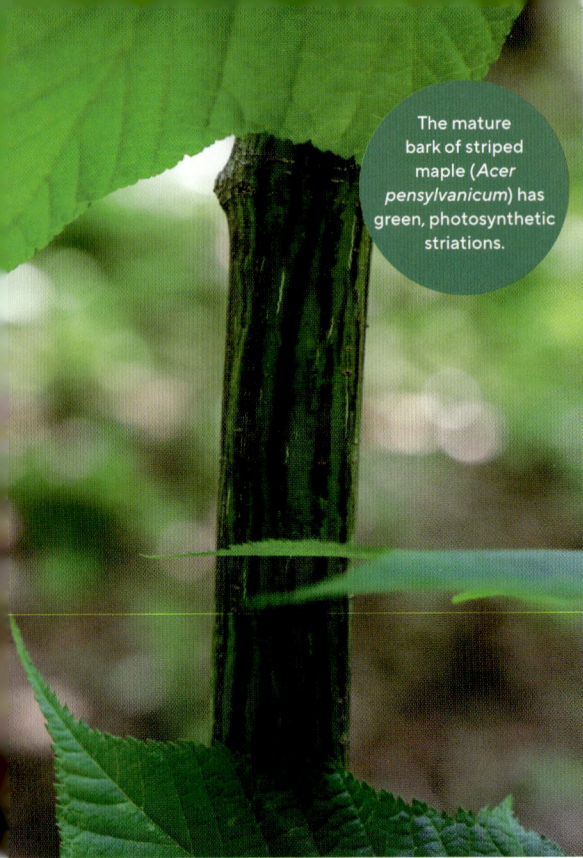

The mature bark of striped maple (*Acer pensylvanicum*) has green, photosynthetic striations.

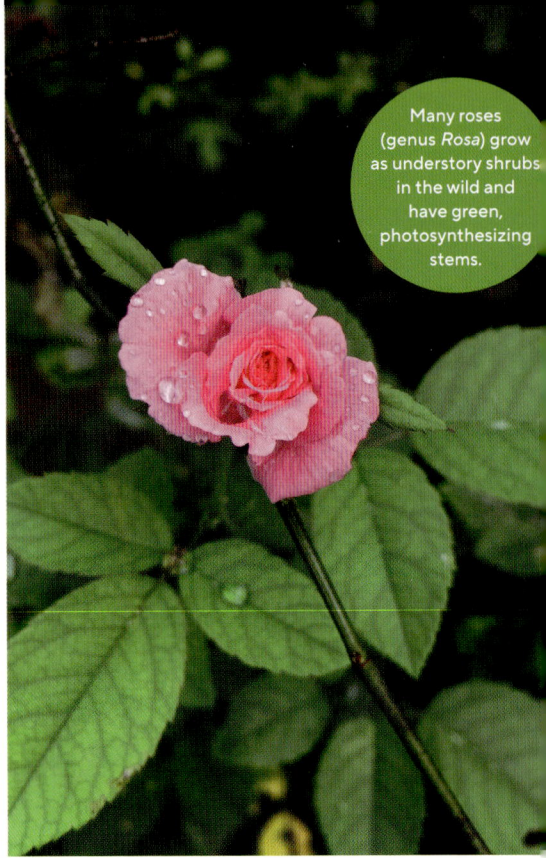

Many roses (genus *Rosa*) grow as understory shrubs in the wild and have green, photosynthesizing stems.

37
Can you find a tree or shrub with a green stem?

A young photosynthetic stem of highbush blueberry (*Vaccinium corymbosum*) is almost as green as a leaf.

On most plants, leaves are the sunlight-harvesting engines of photosynthesis. They have evolved to expose large surface areas to the sun. However, none of the steps in photosynthesis are necessarily unique to the leaf. Any part of the plant that is exposed to carbon dioxide in the air and light from the sun, and that has sufficient water, meets the basic conditions required for photosynthesis. Stems certainly meet these conditions in most plants, and many species perform photosynthesis in their stems.

The easiest way to tell if a stem is photosynthesizing is to look for the color green, which is the primary wavelength reflected by chlorophyll. To perform photosynthesis in a stem, a plant needs to load up the cells in the stem tissue with green chlorophyll molecules. Many thin-barked trees photosynthesize, but you may need to peel back or scratch some of the bark to see the green pigment.

One good place to look for green stems is in shrubs like blueberries, roses, and raspberries. In temperate forests, these understory plants can get a head start on the growing season by photosynthesizing in their stems before any plants (including themselves) have expanded their leaves, casting the forest floor in shade.

38
Can you find where the petals were attached on a fruit?

The ovary is the part of a flower that turns into the fruit (see prompt 2). The ovary is often very small before fertilization of the **ovules**, which are structures that contain unfertilized eggs. Once pollen lands on the flower and fertilizes the ovules, hormonal expression causes the ovary to expand and grow. A mature ovary is called a fruit, and the fruit can be attached to the flower in many different ways.

It is often hard to see ovary position in the young flower without a hand lens, but you can see remnants in the mature fruit. Your kitchen may be an easy place to start. Take a blueberry and a peach. On one end, you will see where the fruit was attached to the stem by a stalk. The other end is what is called the **blossom end**, where the rest of the flower, or blossom, was attached.

A **superior ovary** is one that sits just above the petals, as in this Japanese stewartia (*Stewartia pseudocamellia*).

Sometimes the petals and other floral structures sit on top of the ovary. This is called an **inferior ovary**, as in this possumhaw viburnum (*Viburnum nudum*).

On the blueberry, there is a small crownlike structure at the top. These are remnants of the calyx. You may even be able to find shriveled-up petals sitting on top. This indicates that the blueberry was sitting below the petals and is considered an inferior ovary.

In a peach, the blossom end is pretty smooth, with no noticeable structures at the tip. (There may be a small protrusion at the very top, which is actually the remnant stigma.) This indicates that the peach is a superior ovary; the petals are attached below the fruit.

39

Can you find a fruit with many seeds and another with one seed?

Downy hairs help a broadleaf cattail (*Typha latifolia*) disperse its tiny fruits on the wind.

Though seeds develop within the fruit, they start off as ovules. After pollination, sperm from the pollen grain enters the egg of the ovule, forming a seed. Some plants produce a single seed per fruit, while others produce many.

The size of the seed relates to how much nutrition the mother plant packs away for the next generation. There is usually a tradeoff between size and number of seeds. Those that produce many low-provisioned seeds are **r-selected**, meaning they rely on pure numbers for their next generation to survive. Those that produce few large seeds are **k-selected**, relying mostly on maternal provisioning for survival.

A species' niche and habitat often determine its survival strategy. Those that grow in low-resource conditions are often k-selected. For example, avocados usually grow in the tropical understory, where sunlight is limited. In contrast, the broadleaf cattail (*Typha latifolia*) grows in bright, resource-rich ecosystems and produces nearly a quarter of a million seeds wrapped in tiny, hard fruits.

The pit of an avocado (*Persea americana*) is a single large seed.

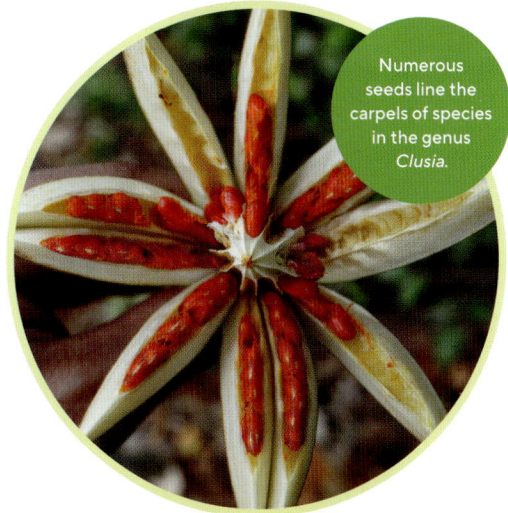

Numerous seeds line the carpels of species in the genus *Clusia*.

Three seeds sit in the open fruit of Chinese bittersweet (*Celastrus orbiculatus*).

40
What are the smallest and largest seeds you can find?

Millions of minuscule seeds are produced from a single orchid flower, and they disperse like dust on the wind. They have very little seed coat. Also, because the nutritive tissue of orchids is sometimes composed of only a single cell carrying essentially zero nutrients, these seeds must make contact with a certain fungus that can harvest and share nutrients to help them germinate and grow. The orchid will repay the fungus by sharing sugar produced through photosynthesis. The symbiotic relationship between orchid and fungus will continue throughout the life of the plant.

parts

While you will certainly find some large seeds in your area, you may not find the largest on Earth. That record belongs to the coco de mer palm tree (*Lodoicea maldivica*) seed, which can weigh more than 50 pounds. The smallest seeds belong to orchids and weigh only a few hundred millionths of an ounce. That means the largest seeds are roughly 20 billion times heavier than the smallest seeds! In spite of this astonishing range, all seeds contain the same three basic parts—the embryo of a new plant, some nutrition provided by its parents, and a protective coat surrounding these contents. All variation in seed size arises from tweaking these three elements.

Large seeds like that of coco de mer (*Lodoicea maldivica*) are packed full of nutrients and usually have thicker seed coats to protect those precious (and delicious) nutrients from hungry animals. Large, well-provisioned seeds like this one are typical of environments where light or other plant needs are in high demand. In tropical understories, for instance, young plants struggle to harvest sunlight, so a large stock of nutrients in their seeds allows them to start growing toward the brighter levels above.

41

Compare the placement of seeds inside a tomato and a cucumber.

Seeds are not arranged the same way from one fruit to another.

Cut open a tomato and you'll notice that the seeds are attached along a main central axis down the center. On the other hand, cucumbers hold their seeds along the outer walls of the fruit. The orientation of ovules within a fruit is called **placentation**. Tomatoes have axile placentation, while cucumbers have parietal placentation.

The proper placement of ovules within the maternal body is important. For instance, if an embryo is embedded ectopically, it may not develop. This is true among animals as well as plants. However, what is also true is that ovules are not developed in the same exact place among different species. Most species have axile

placentation like the tomato, while parietal placentation (that of the cucumber) is a close second. There are also several less common arrangements.

This raises the still unanswered question of why different placentation types exist. Interestingly, species with parietal or axile placentation tend to have many seeds per fruit, while fruits with few seeds tend to have other arrangements. Seeds have evolved many shapes and sizes across plant history—consider the tiny seeds of an orchid and the larger seeds of a pumpkin. As seed shape and size change over time, it may be that plants also need to modify how they pack their seeds within the fruit.

A sliced tomato clearly shows its axile placentation, with seeds attached to a central axis.

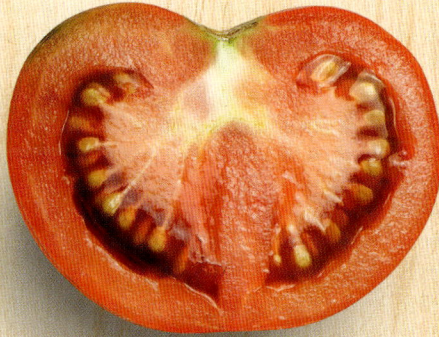

A sliced cucumber shows seeds attached to the outer walls of the fruit.

42

Can you find the portion of a seedling where the stem and the root meet?

The Hypocotyl

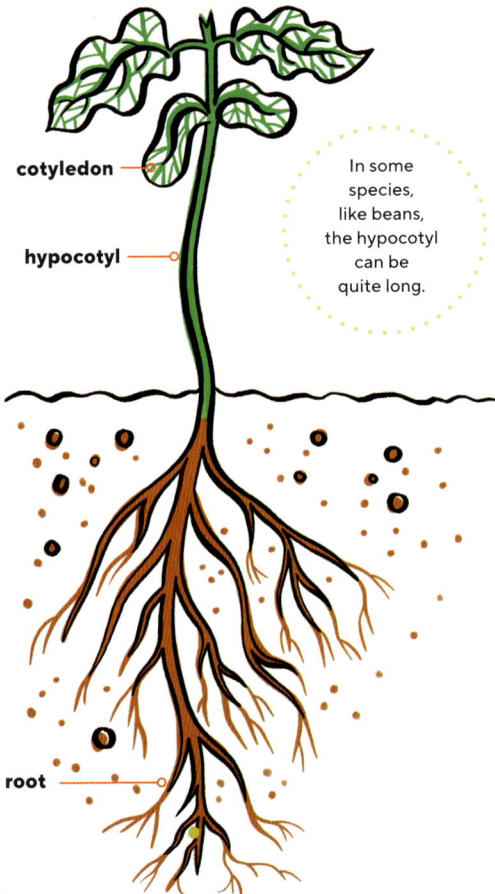

cotyledon

hypocotyl

In some species, like beans, the hypocotyl can be quite long.

root

Look closely at a beet or a bean sprout the next time you eat one. Is this a stem or a root? You might be surprised! There is an intermediary zone between the root and the stem system, a middle ground where the two organ systems meet. This is called the **hypocotyl**. In a germinating seed, this is the ambiguous-looking region just below the cotyledons.

The hypocotyl is a structure that has been given a formal name and definition even though it may not have a consistent or clear function. After all, the plant does not care what we call its organs. The hypocotyl may simply be the necessary space between two structures.

Within the frequently studied flowering plant species *Arabidopsis thaliana* (thale cress, a small plant from the mustard family), there is evidence that light-driven move-ment of the hypocotyl may help push the young seedling through the soil or help with placing cotyledons and early leaves (see prompt 11) in the light's path during the early stages of gemination. On the other hand, in beets, the hypocotyl has been modified into a fleshy storage organ—which we harvest and eat in our salads.

leaf

In mature beets, the hypocotyl has been modified into a fleshy storage organ.

hypocotyl

taproot

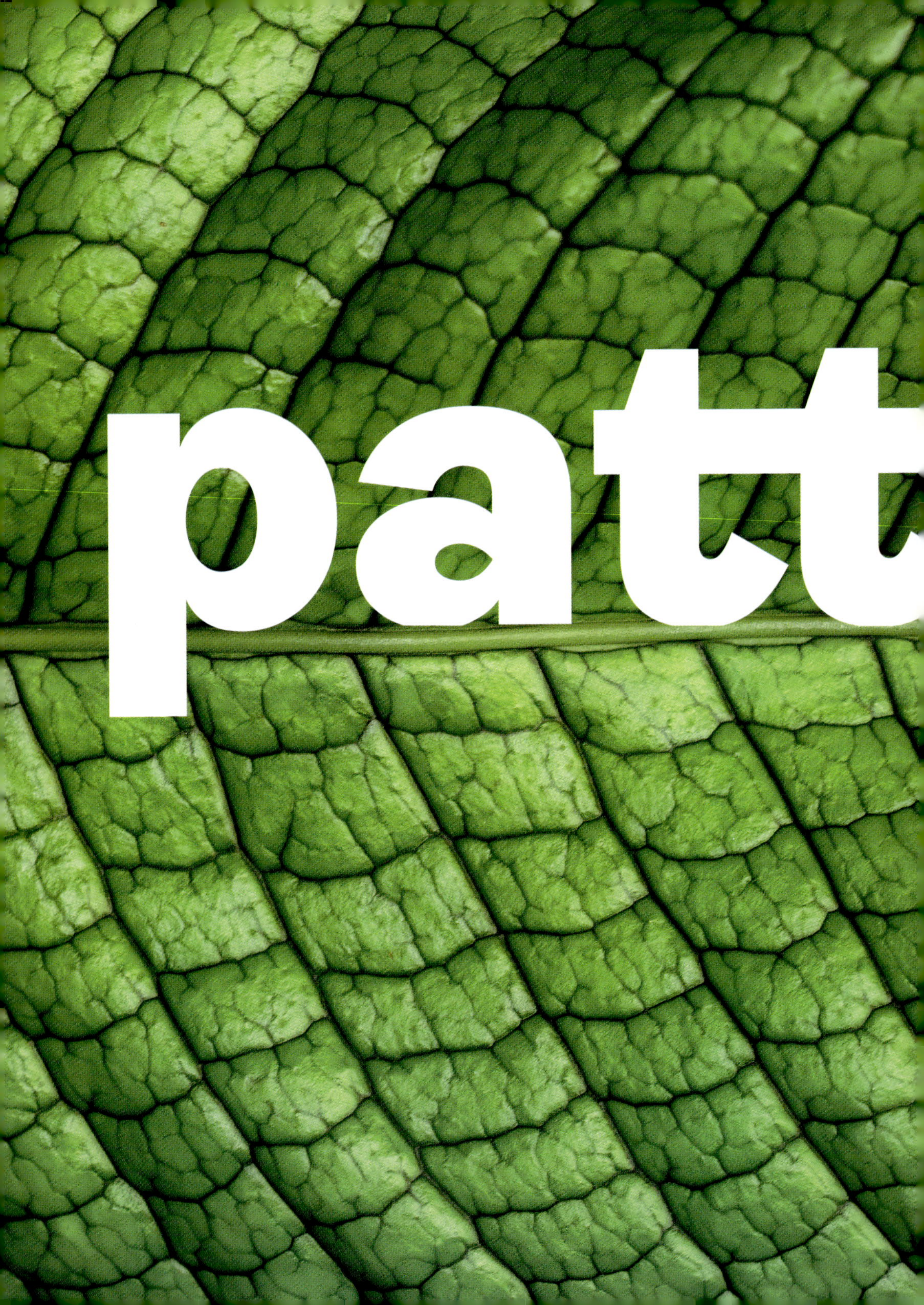
patt

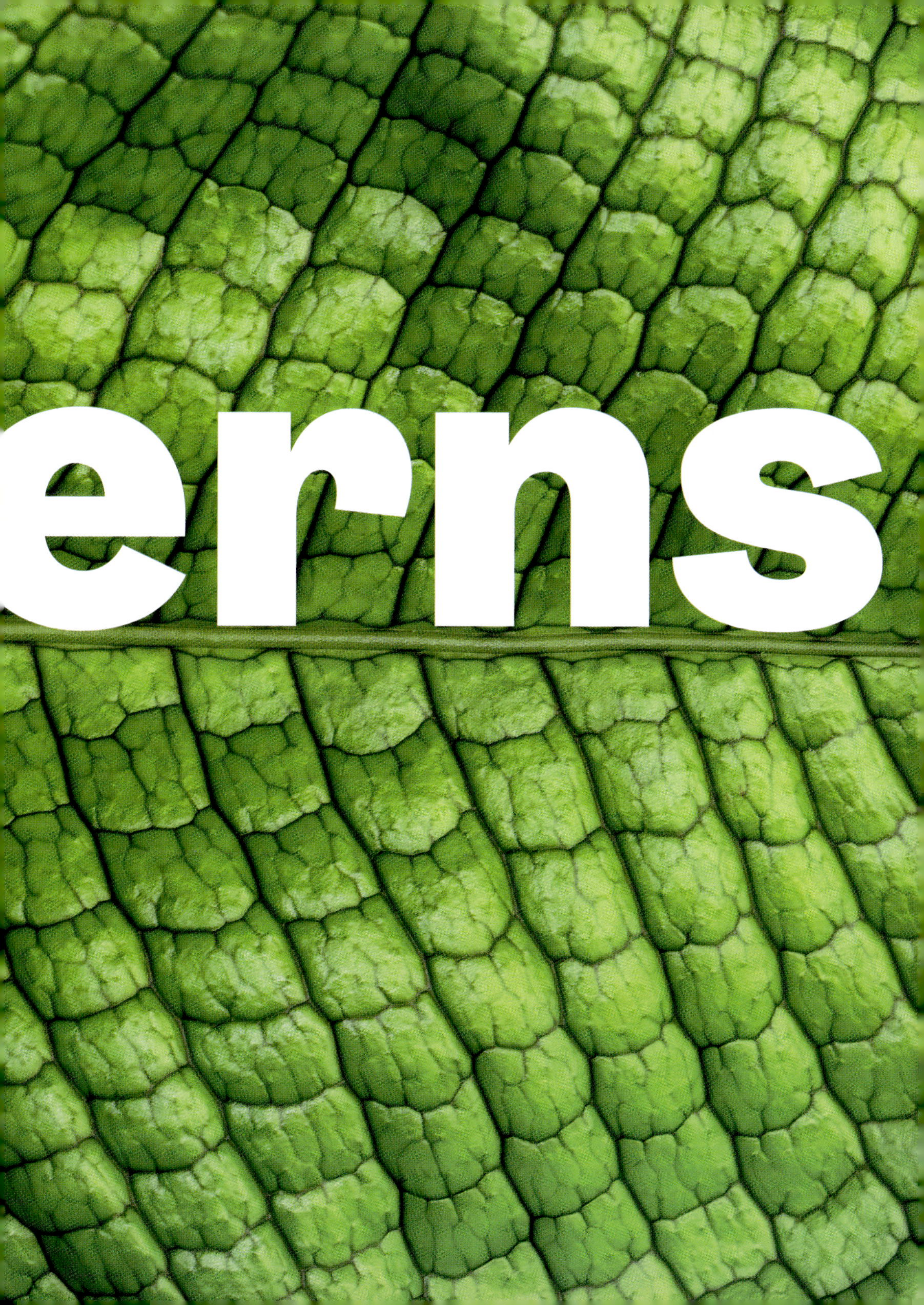
erns

Sugar maple
(*Acer saccharum*)

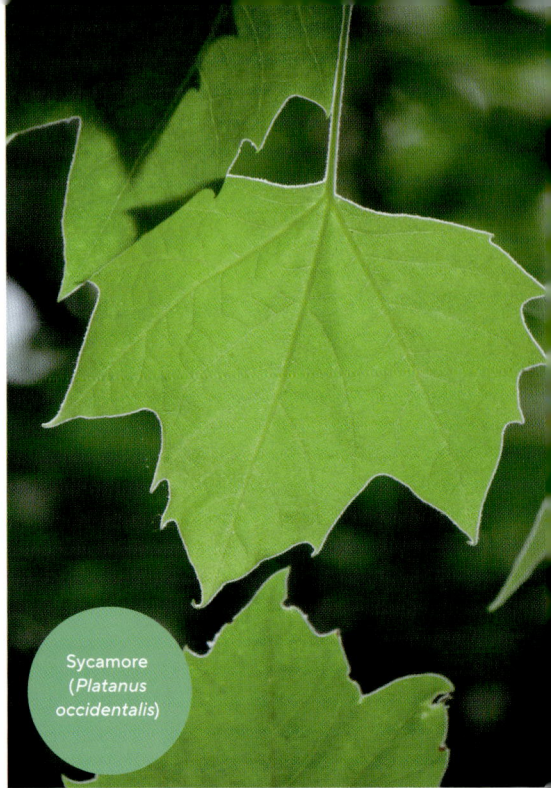

Sycamore
(*Platanus occidentalis*)

Lotus
(*Nelumbo nucifera*)

43
Can you find leaves of plants that you think may be closely related?

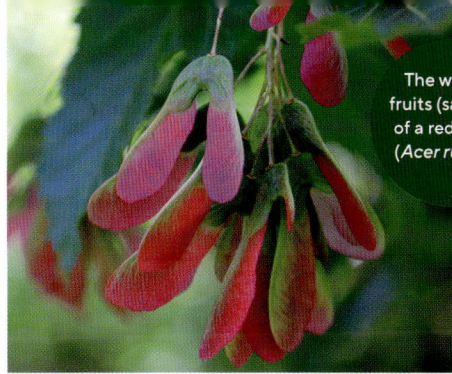

The winged fruits (samaras) of a red maple (*Acer rubrum*)

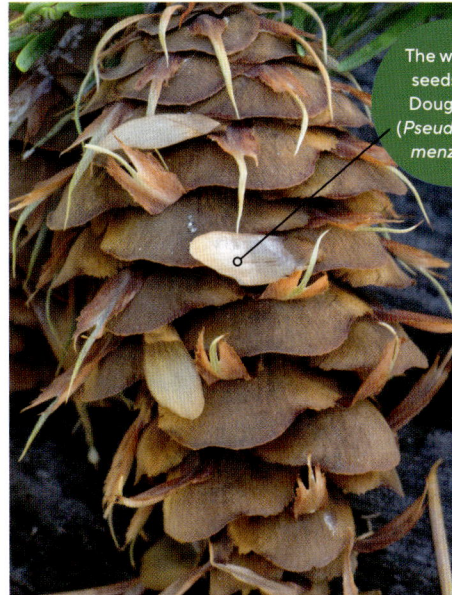

The winged seeds of a Douglas fir (*Pseudotsuga menziesii*)

Take a look at the photos at left.

You'd probably assume the maple and the sycamore are close relatives, right? Not quite.

You may resemble your cousin more closely than you resemble a randomly chosen classmate or coworker, and that's because you and your cousin are related. When we speak about being related to someone, we mean that we share ancestors in our recent past. This same logic is true when we look across species on timescales of millions of years. For example, if you look at a dog, a cat, and a goldfish, the shared traits of fur, four bony legs, and similar teeth reflect the fact that dogs and cats are more closely related to each other than either is to a goldfish. However, sometimes this logic based on appearances can be misleading, especially in plants. In some cases, the appearance of two unrelated plants can converge.

Surprisingly, among the photos at left, it is actually the sycamore and lotus that share a more recent common ancestor. More recent shared ancestry results in genetic code with greater similarity but not always in more similar physical appearance.

Sometimes, small genetic or developmental changes can lead to dramatic differences in physical traits. For example, growing as a tree or as an herbaceous plant requires very few genetic tweaks. Very distantly related lineages also evolve to resemble one another, often in response to similar environmental conditions. When two different evolutionary paths arrive at a similar-appearing destination, usually to solve a similar problem, we call this **convergent evolution**.

Maple and pine trees offer an example of convergent evolution in the structure of their fruits and seeds, respectively. Pines are not flowering plants, so they do not have fruits. The seeds of a pine tree are instead held inside the pine cone, lying flat on top of each woody scale. The pine seed structure resembles that of a maple fruit. (Maple fruits are called **samaras** and contain the seed within.) Both pine seeds and maple fruits have converged on a similar strategy to be carried off by the wind on their wings, but they came from very different starting points.

patterns

44

Find a flower that is open when no others are.

Some bees and insects are only active early in the spring weeks, as ephemeral as bloodroot (*Sanguinaria canadensis*) and trillium (*Trillium*).

When you think of flowers,

you may imagine warm spring days or hot summer months. Indeed, many plants flower when it is warm out and insect pollinators are abundant. When plants are actively photosynthesizing in the summer, a steady supply of sugar is readily available from the current year's leaves to build flowers, fruits, and seeds. However, there is an issue with flowering in peak summer: It's crowded.

Leaves on trees make it difficult for wind-pollinated plants to move their pollen, and insect pollinators have many options to choose from. To get around these issues, some plants flower early in the growing season. Wind-pollinated trees, like some maples and hazelnuts, bloom early to disperse their pollen far distances without being impeded by the leafy canopy. Insect-pollinated plants like cherries and ephemeral spring flowers like bloodroot will flower before the first day of spring.

Many of the plants that flower early do so before they even leaf out. This pattern is called **hysteranthy** (or **proteranthy**) and is contrasted with **synanthy**, which is when leaves and flowers appear simultaneously. Flowering early before leaf-out helps catch those hungry early pollinators who haven't foraged nectar all winter. In fact, some bees and insects are active only in early spring and depend on these first blooms.

A Japanese cornelian cherry (*Cornus officinalis*) blooms in early spring before leaf-out.

Ozark witch hazel (*Hamamelis vernalis*) flowers in the middle of winter.

patterns

105

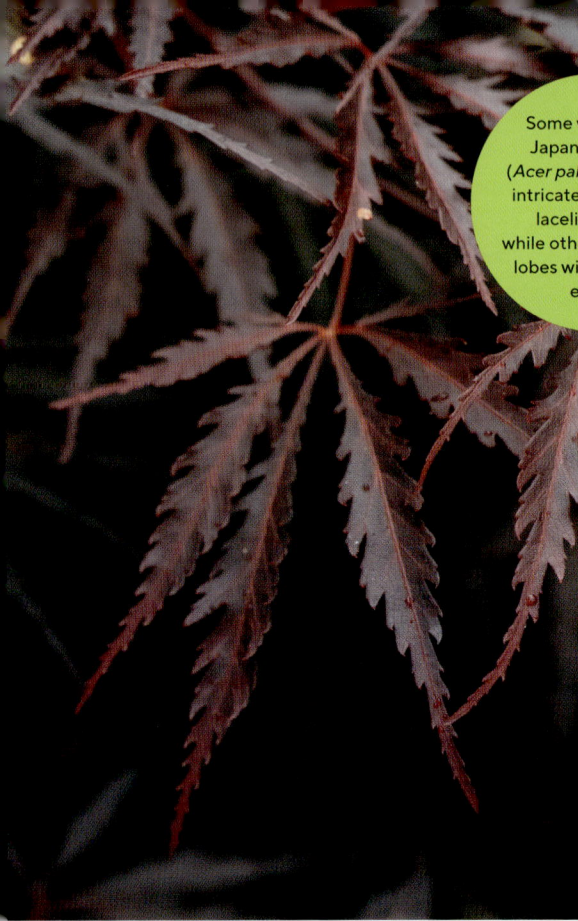

Some varieties of Japanese maple (*Acer palmatum*) have intricately dissected, lacelike leaves, while others have fuller lobes with smoother edges.

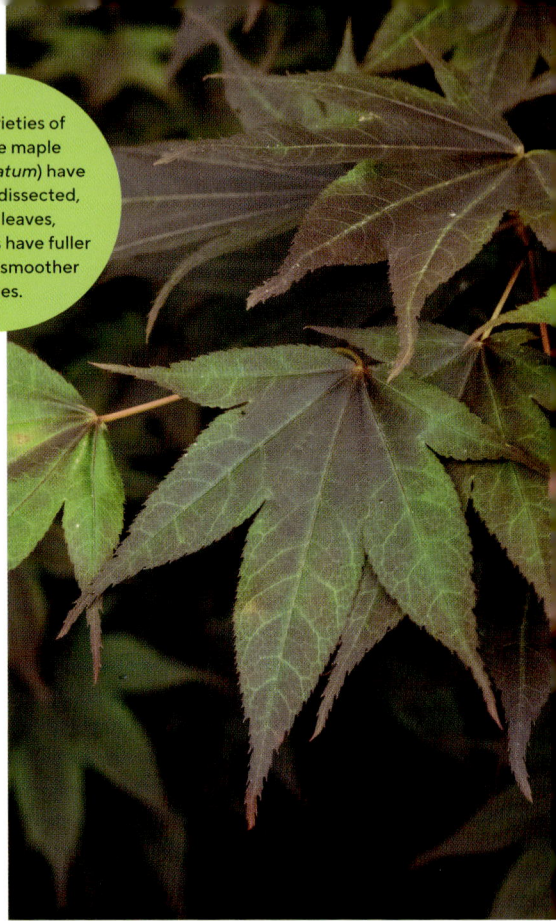

45
Compare two leaves on different individuals of the same species.

Resemblance is how we can tell (or at least hypothesize) that two different individuals belong to the same species. However, that doesn't mean that all individuals of one species look identical. In fact, high variation among individuals within a species can be on full display in the garden. Selective breeding has been used to generate dramatically different-looking varieties of the same species.

When varieties maintain their distinctive characteristics even when they are grown near each other in the same environment, we know that their differences are explained by genetics.

In contrast, **phenotypic plasticity** is another potential cause of variation between individuals of the same species. A phenotype is a characteristic, or trait, of an organism, and *plasticity* means the ability to change or vary. We see phenotypic plasticity at work when individuals develop different characteristics in response to different environments, not because of genetic differences. An example is in hydrangea.

The showy parts of hydrangea flowers are typically not the petals but colorful **bracts**. In bigleaf hydrangea (*Hydrangea macrophylla*) these modified leaves turn blue when growing in more acidic soils and pink when growing in soils that are more alkaline. The change in flower color is due entirely to the soil conditions, and the same individual can produce bracts of several different colors. This happens when roots feeding one part of the plant are tapped into a spot of soil that has a different pH compared to roots in another patch of soil.

The color variation in bigleaf hydrangea is related to soil conditions, not genetics—an example of phenotypic plasticity.

46

What is the most brightly colored plant part you can find that is not a flower or fruit?

All colors in the biological world come from either chemical pigments or structural traits. The reds, greens, and yellows of plant leaves and flowers are typically due to chemical pigments.

We tend to associate color in plants with floral organs, which are often pigmented: anthocyanins make reds, carotenoids make yellows, and a mix of them makes browns and purples. However, flowers and fruits are not the only plant parts that produce colors. In fact, some of the brightest colors come from young developing leaves, which can be jam-packed with anthocyanins.

The color blue in nature is a fascinating phenomenon. Unlike the dye in your favorite pair of jeans, the stunning metallic blue color of many plants (as well as birds and butterflies) is often not from pigment. Rather, this color arises from cell structures that selectively absorb and reflect light.

Next time you are looking at a plant, try to appreciate the color outside of the flower.

TOP: The leaves of the tropical *Cespedesia spathulata* are bright red.

MIDDLE: Even fern leaves, like those of this young and aptly named pink maidenhair (*Adiantum macrophyllum*), can be red.

BOTTOM: There are several ways that plants have evolved structurally to reflect blue light. One is the inclusion of spiral structures within the cell walls of the epidermis (as in this *Microsorum thailandicum*).

patterns

47

Can you find a plant growing on another plant?

Large plants create entirely new habitats for other organisms to live on, including other plants. When a plant grows on top of another plant without any direct connection to the ground, we call it an **epiphyte**. The word *epiphyte* is derived from the Greek words *epi,* meaning "upon," and *phyton,* meaning "plant"—literally, "upon another plant." We give these plants a special name because living as an epiphyte poses some unique ecological challenges.

Since epiphytes have no direct connection to the soil, they lack reliable access to water and nutrients. Even in the wettest climates like tropical rainforests, life as an epiphyte can be quite dry and harsh. Soil on the ground is like a sponge, holding water in place for plant roots to soak up at their own pace. Epiphytes must rely directly on more fleeting water sources like fog and rain.

In spite of the challenges of living on another plant, many lineages of plants have evolved epiphytic lifestyles, especially the orchids, bromeliads, mosses, and ferns. This suggests that the benefits of escaping a crowded understory and reaching higher into the canopy can outweigh the challenges of being an epiphyte. These benefits include more access to sunlight, better visibility to pollinators, or a higher perch for releasing spores into the breeze.

A bromeliad makes its home on the branch of a tree among other epiphytes.

This *Tillandsia* absorbs water directly into its leaves, not just through the roots, which enables it to drink condensed moisture right out of the air.

Some epiphytic plants, like this staghorn fern (*Platycerium*), grow into basketlike shapes that catch falling leaves and other debris, creating their own composting soil factories in the canopy!

In the temperate region, epiphytes are more likely to be mosses, lichens, and liverworts. This scalewort (*Frullania*) is a type of liverwort.

An epiphytic orchid has long roots that can absorb water that collects on the bark or in the crotches of a tree.

48
Can you find a flower that does not produce petals?

Whether we are plucking them off to determine if someone loves us or sprinkling them on the table for a fancy dinner after we realize they do, petals are the most recognizable parts of a flower. In many flowering plants, petals are large, showy, and colorful structures that aid in attracting pollinators like bees, butterflies, and birds.

However, not all flowering plants use animals for pollination. Many use the wind to move pollen from flower to flower. When a plant evolves to do this, they tend to lose their petals over evolutionary time. Why invest in showy petals if they are not trying to attract animal pollinators? Petals are costly—they are large, pigment-filled structures that tend to use large amounts of water and nutrients. This evolutionary reduction has happened multiple times across the flowering plant tree of life including in the oaks, birches, and sycamores.

Instead, wind-pollinated plants tend to invest in copious amounts of pollen and large, feathery stigmas. This is because wind pollination is very inefficient. What are the odds that a pollen grain 6,000 times smaller than a grain of rice will land on a flower a mile away?

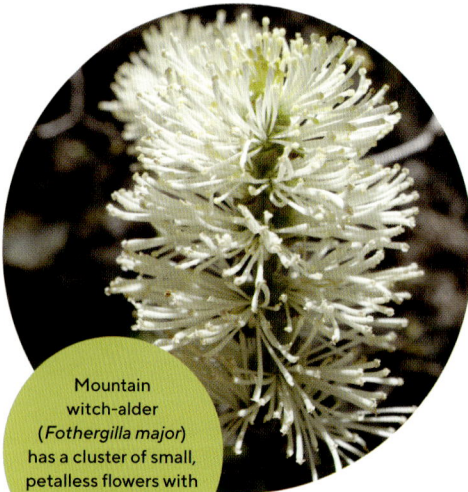

Mountain witch-alder (*Fothergilla major*) has a cluster of small, petalless flowers with showy stamens.

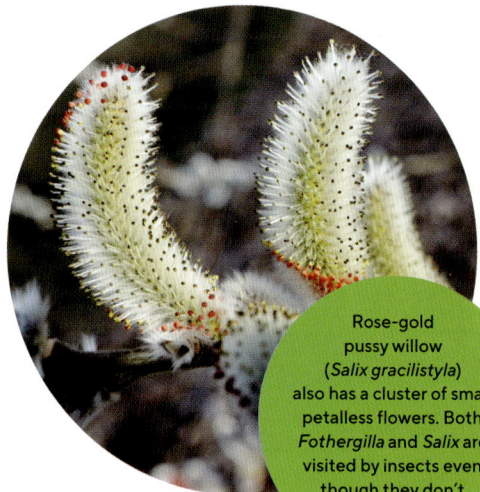

Rose-gold pussy willow (*Salix gracilistyla*) also has a cluster of small petalless flowers. Both *Fothergilla* and *Salix* are visited by insects even though they don't have petals.

Perennial glasswort (*Salicornia ambigua*) has numerous petalless flowers containing large white stigmas.

EARLY BLOOMERS

Wind-pollinated plants often open their minute petalless flowers early in the spring or dry season, when no leaves are on the trees, in order to help pollen travel far distances unimpeded (see prompt 44).

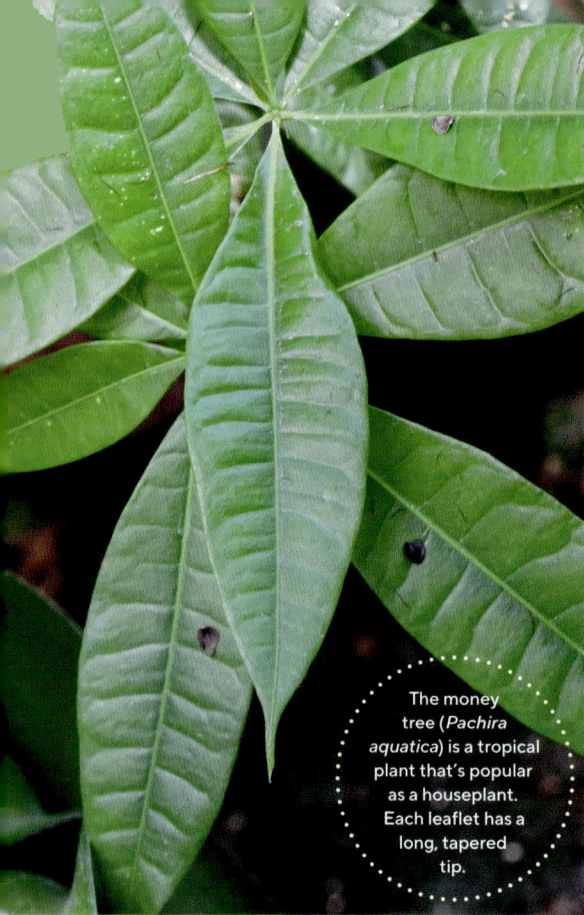

The money tree (*Pachira aquatica*) is a tropical plant that's popular as a houseplant. Each leaflet has a long, tapered tip.

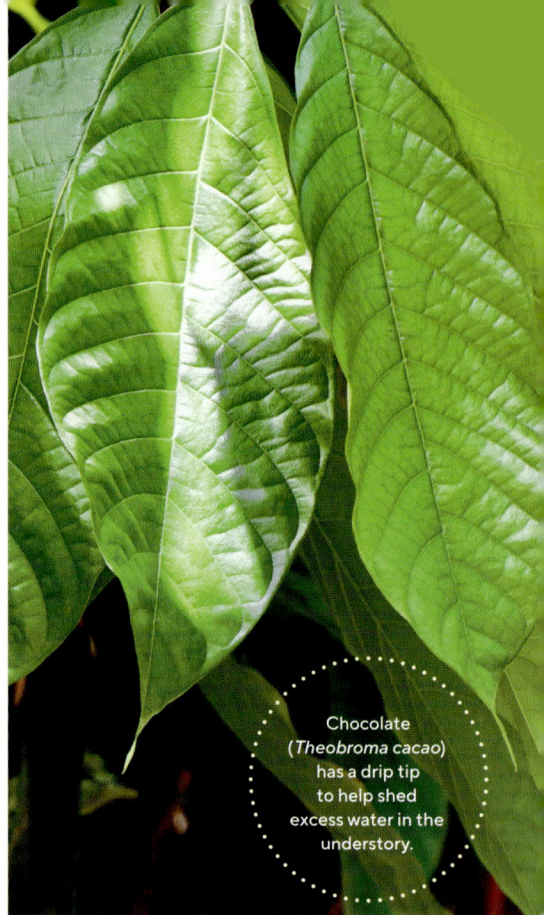

Chocolate (*Theobroma cacao*) has a drip tip to help shed excess water in the understory.

49

Can you find a plant with long, tapered leaf tips?

HINT: If you are in the temperate region, look at your houseplants.

A rubber tree (*Hevea brasiliensis*) is another tropical plant with a slender drip tip.

If you live in the tropics (or have tropical houseplants), you might observe that many plants have a particular kind of leaf tip. The tips of the leaves will be long and **acuminate**, tapering to a point. These points are called drip tips. The name reflects their function: As rain drops onto the leaf surface, it smoothly beads up and rolls off down the tip. This is different from straplike leaves of grasses or tulips, which don't end in a long tip and are pointed upward to help funnel water.

Why don't leaves want to have a wet surface? Leaves need to be clean and clear from debris to capture light and photosynthesize. In tropical environments, any wet surface can be a habitat for microorganisms like algae or fungi, or even larger organisms like mosses and leafy liverworts. If a leaf stays wet, these organisms can colonize it, smothering its upper surface. This can reduce light availability, decrease gas exchange, and potentially harbor pathogens. Drip tips are thought to be an adaptation to shed water more efficiently to help tropical leaves stay healthy.

50
Look for a mossy habitat.

What characteristics does it have?

Mosses are an extremely diverse lineage of plants, with around 12,000 described species (see prompt 31), and yet they tend to grow in a small number of predictable habitats. We often find moss growing in places with little to no soil, like the surface of a rock or the bark of a tree. We also expect to find mosses in shady, ever-wet sites like on the banks of a creek or on tree branches in a rainforest. In human-built landscapes, moss is a fixture of sidewalk cracks, brick walls, and old wooden fences.

Mosses don't produce true roots. Rather, they make simpler, short, rootlike

Mosses that have adapted to periodically dry habitats spread open their green leaves when wet.

Crisped pincushion moss (*Ulota crispa*) grows on a rock.

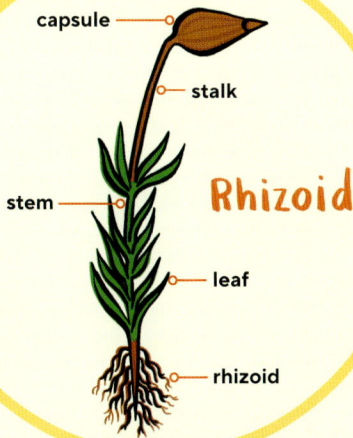

capsule

stalk

stem

Rhizoid

leaf

rhizoid

A species of live sheet moss (*Hypnum*) grows on a damp shady log in forest understory.

threads called **rhizoids**. Rhizoids primarily serve to anchor mosses to their substrate, and while they do help absorb water, they access only the very surface layer of soil, rock, or bark. Since they cannot access water deep in the soil like a plant with a larger root system, most mosses have evolved to survive cycles of drying out and springing back to green life when water arrives.

Hence, we expect to find mosses growing in places where other plants would dry out and die, like on rocks and tree bark, or in places with constant wetness, like bogs and riversides.

patterns

51
What is the tallest moss you can find?

Not every plant has a true vascular system. In fact, the first plants to migrate onto land lacked these water-conducting tissues, and the modern mosses, liverworts, and hornworts also do not produce xylem or phloem. Their inability to move large amounts of water and nutrients through a set of vascular highways is one reason why they remain relatively small.

Some mosses, however, have independently evolved structures very similar to xylem and phloem, called **hydroids** and **leptoids**. Hydroids move water, and leptoids transport sugar. Like tissues in the vascular plants, hydroids and leptoids have helped some moss grow larger.

Boulder broom moss (*Dicranum fulvum*) lacks hydroids and stays low to the growing surface.

A common haircap moss (*Polytrichum commune*) grows almost as tall as neighboring seedlings.

HOW PLANT VEINS CHANGED THE WORLD

Vascular tissues first evolved about 420 million years ago and allowed plants to move much more water and sugar through their bodies (see prompt 1). Veins also provided structural support and allowed plants to grow taller than ever before. These adaptations allowed plants to photosynthesize more, which sucked much more carbon dioxide out of the atmosphere and created more oxygen—changing life on our planet forever!

patterns

The tallest moss ever recorded, *Dawsonia superba*, possesses hydroid and leptoid tissues and reaches nearly 2 feet (60 cm) in height! It is found only in Tasmania.

52

Can you guess how a flower is pollinated based on its appearance?

Hummingbird-pollinated flowers are typically red and have no landing platform, since hummingbirds hover while they feed. These flowers are often long and tubular, producing nectar at their base where only the long beak and tongue of a hummingbird can reach it.

When it comes time to reproduce, plants can't move to find their mates. Many flowering plants have overcome this challenge by evolving mechanisms to entice animals into assisting their reproduction (see prompt 8). Animals can carry pollen from one flower to another and can visit many plants in one day. From the plant's perspective, reproductive success typically depends on pollen moving between different individuals of the same species. But from the pollinator's perspective (imagine a bee or hummingbird), a successful foraging bout could include visits to any number of different plant species, so long as they offer a worthwhile food reward.

By having flowers that appeal to a specific type of pollinating animal (for example, a bee but not a hummingbird), a plant can reduce the likelihood that its pollen will be wasted by being deposited on a different species' flower. This has resulted in the evolution of so-called pollination syndromes—suites of floral traits associated with attracting a specific type of pollinator based on that pollinator's sensory biases and physical and behavioral traits.

TOP: Purple-flowering raspberry (*Rubus odoratus*) has pinkish-purple flowers with wide petals that serve as a large landing platform for bees, as well as wasps and other small insects.

MIDDLE: Three-leaved rattlesnake root (*Nabalus trifoliolatus*) has downward-facing inflorescences with elongated parts that bees like this bumblebee can grasp.

BOTTOM: Swamp azalea (*Rhododendron viscosum*) has long, tubular, white flowers that attract hawk moths, butterflies, and bees.

53

Can you find both a dry fruit and a fleshy fruit that open when ripe?

The structure of a fruit is usually related to its mode of dispersal. For instance, large fleshy fruits with a lot of nutritional investment, like avocados, are usually dispersed by vertebrate animals (see prompt 70). Small, dry fruits with less nutritional tissue, like those of maples, are often dispersed by the wind. Some dry fruits, like milkweed and cottonwood, **dehisce** or open up, releasing individual seeds that have wings or tufts of hair (see prompt 13). These

The dry, dehiscent fruit of butterfly milkweed (*Asclepias tuberosa*) opens to reveal seeds with long tufts of white hair that catch the wind.

seeds are swept up by the air currents and carried far from the mother plant, eventually germinating to produce the next generation.

However, while most dehiscent fruits are dry, some are fleshy, and these tend to be dispersed by animals. Nearly all fleshy dehiscent fruits have seeds with coatings called **arils**. Arils are usually brightly colored to help entice animals, often birds, to swoop down and nibble on a few of the seeds.

Korean spindle tree (*Euonymus oxyphyllus*) bears colorful fleshy fruits that open when ripe, revealing red arils that easily catch the eye of passing birds.

Pomegranates (*Punica granatum*) are fleshy and dehiscent. We usually purchase these fruits when they have not yet reached their full natural maturity. In nature, the pomegranate stays on the tree until it splits open along its seams—truly bursting out of its own skin. As it opens, it reveals a beautiful matrix of seeds covered in juicy red arils—the part we eat.

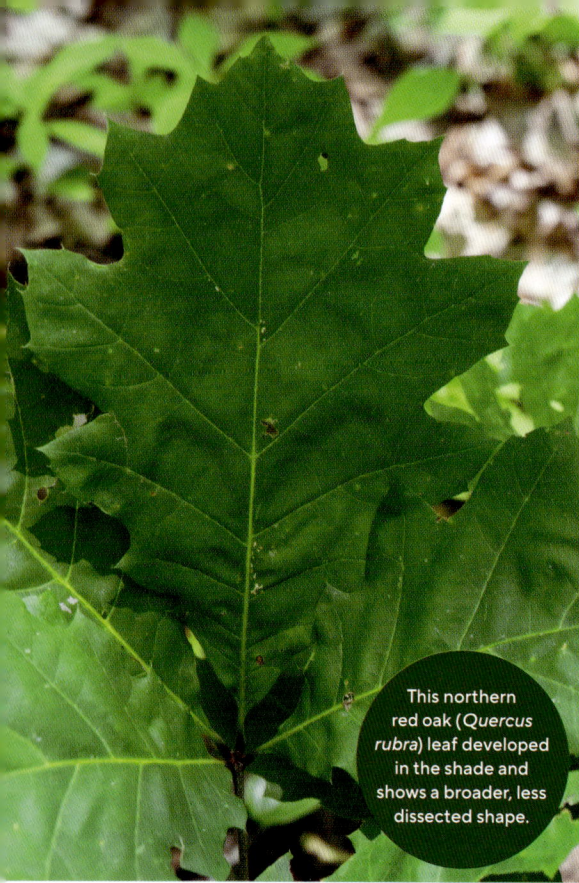

This northern red oak (*Quercus rubra*) leaf developed in the shade and shows a broader, less dissected shape.

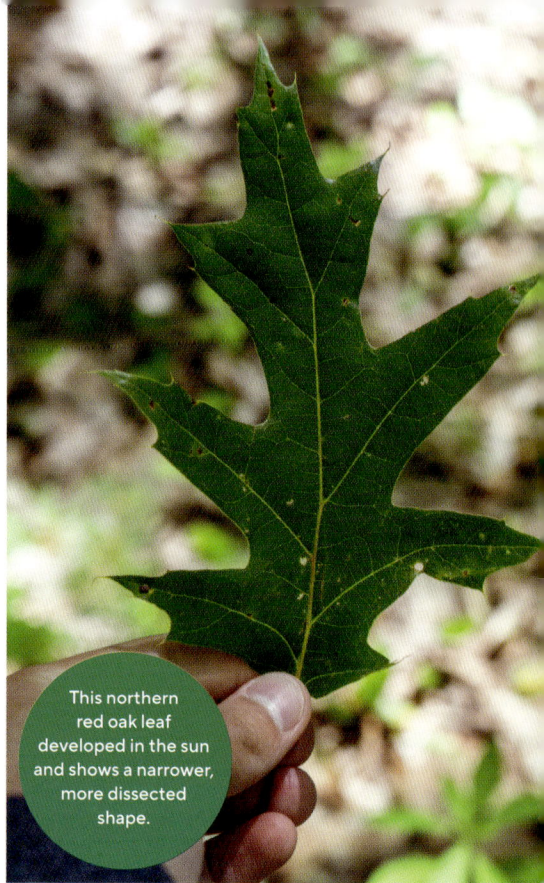

This northern red oak leaf developed in the sun and shows a narrower, more dissected shape.

54
Can you find two different-looking leaves on the same individual plant?

Passionflowers (*Passiflora*) may have evolved such widely varying leaf shapes to make them more difficult for butterflies in the genus *Heliconius* to recognize. When these butterflies lay their eggs on passionflowers, the resulting caterpillars become the plants' primary herbivore threat.

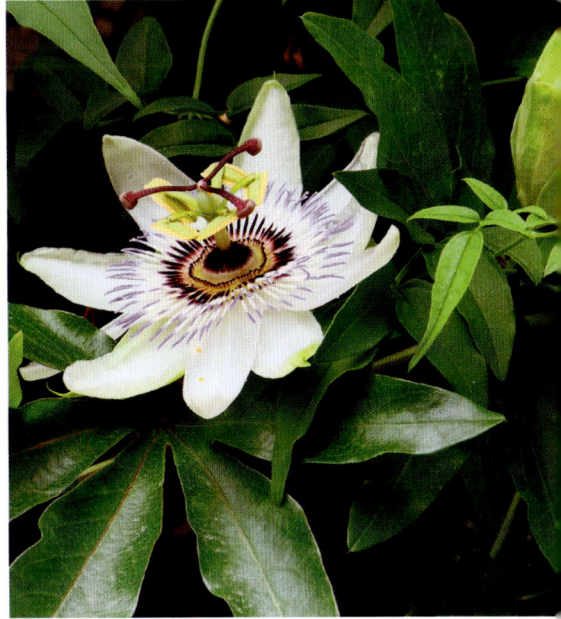

In many species of mulberry, like this white mulberry (*Morus alba*), no two leaves along a single branch look the same.

We usually think of a plant species as having one characteristic leaf shape. Subtle differences in leaf shape, like the size of teeth or the angle of lobes, are often invoked as useful traits for distinguishing among closely related species. (For exceptions, see prompt 43.) However, in some plants, leaf shape is far from consistent and can differ significantly within a single individual. This phenomenon is called **heterophylly**, from the Greek for "different leaves."

Oaks (*Quercus*) are one familiar group of plants that often exhibit heterophylly. In some oak species, the leaves in the canopy and on the outer tips of branches are more deeply lobed and delicate than leaves produced lower on the tree and closer to the trunk. In this case, heterophylly is a response to the amount of sunlight hitting each individual leaf, leading to the production of "sun leaves" and "shade leaves."

Another fascinating example of heterophylly can be found in the passionflowers (*Passiflora*). In some species, leaves change dramatically in shape as they age. You could pick a leaf from two different parts of the same plant and never imagine that they came from the same species, let alone the same individual.

patterns

125

55
What is the smallest repeating unit you can find in a sunflower?

HINT: Look closely at the center.

Sunflowers, especially domesticated varieties, appear to produce very large flowers. However, this is a trick. What we perceive as a single flower is a dense and highly coordinated inflorescence. A single sunflower may contain more than 1,000 individual flowers or florets, each of which will give rise to one sunflower "seed" (actually a single-seeded fruit called an achene).

If you look closely at a sunflower, you will notice that the hundreds of individual constituent flowers come in two varieties. The flowers in the center are called **disc florets**, while those on the outer edge are called **ray florets**. The disc florets are packed densely in the center where they await pollination, while the ray florets around the outside each produce one large petal. Together, the ray florets create one continuous ring of petals, forming a large, attractive target for potential pollinators. This structure is called a **capitulum** and is characteristic of the entire sunflower family (Asteraceae). This strategy is evidently very successful from an evolutionary standpoint, as the Asteraceae has diversified into a family of more than 30,000 described species, one of the largest among the angiosperms.

The Jerusalem artichoke (*Helianthus tuberosus*) is in the sunflower family. Its characteristic inflorescence shows a ring of disc florets with small petals surrounded by a ring of ray florets, each with one large petal.

The Canada goldenrod (*Solidago canadensis*) is also in the sunflower family. Its bloom consists of many small inflorescences of disc and ray florets.

Hundreds of individual disc florets make up the center of a sunflower.

Round-lobed hepatica (*Hepatica americana*) show a mottled pattern of two different shades of green. This is a common strategy among ephemerals that stand out in the early spring, before other plants have leafed out.

56
Can you find a mottled or camouflaged leaf?

HINT: Look in the early spring or winter if you are in the temperate region.

Little brown jug (*Hexastylis arifolia*)

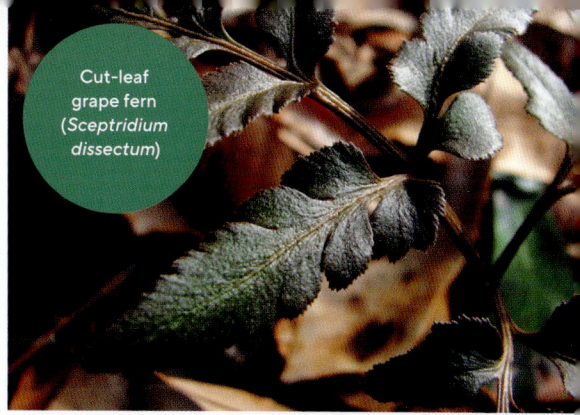
Cut-leaf grape fern (*Sceptridium dissectum*)

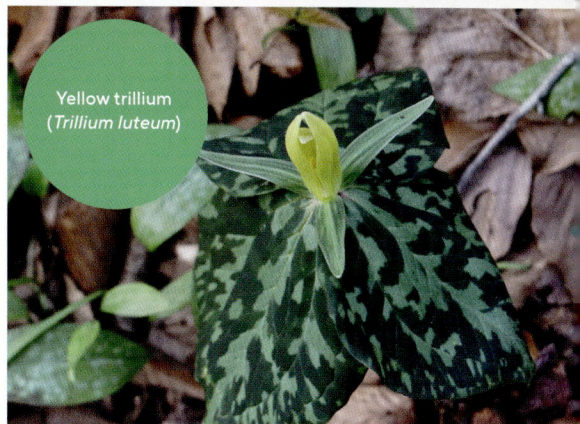
Yellow trillium (*Trillium luteum*)

Camouflage isn't just for animals.

Many plants need to hide from herbivores. Some have devised interesting ways to do so. By changing leaf color, evergreen species like Boston ivy (*Parthenocissus tricuspidata*), little brown jug (*Hexastylis arifolia*), and cut-leaf grape fern (*Sceptridium dissectum*) can hide in the leaf litter. During the growing season, leaves will often be bright green, amplifying chlorophyll to capture as much sun as they can in the dense understory.

As fall passes and deciduous trees drop their leaves, the canopy opens, beaming light down onto the forest floor. However, now that all the vegetation is gone, the leaves of understory evergreen herbs are exposed, ready to be eaten by herbivores like deer. To camouflage themselves from predators, these species will often upregulate pigments like anthocyanins in the leaves to mottle or darken the foliage. By hiding in plain sight, they may be able to avoid predation and keep their leaves active for longer (see prompt 63).

patterns

Ghost plant (*Monotropa uniflora*) is a parasitic plant that can be almost pure white, with rudimentary, scalelike leaves. It siphons all its sugar from its host through a fungal partner.

57
Can you find a plant with no green parts?

We define plants by their colors.
Green is by far the most conspicuous. Plants are green because they produce chlorophyll, the pigments that capture sunlight and help drive photosynthesis. These pigments appear green to us because they absorb visible light in the blue and red wavelengths and reflect green wavelengths. Plants that photosynthesize are called **autotrophs**, meaning they produce their own food (as opposed to humans and other animals that are **heterotrophs** and must eat food to survive).

However, chlorophyll and its associated photosynthetic machinery is expensive to make. It requires a large amount of nitrogen and magnesium, which are important nutrients for plants. Some plants have figured out how to forgo these expensive molecules altogether by becoming parasites. Parasitic plants tap into the reserves of their host and steal valuable carbon, water, and nutrients. This allows them to live without photosynthesizing, and since they don't photosynthesize, they don't require large green leaves.

Many plants across the tree of life have evolved parasitism. Some of them have gone at full throttle and lost the ability to produce chlorophyll altogether, while others, like the mistletoes (Loranthaceae) and paintbrush plant (*Castilleja*), still produce some of their own sugar.

Pinyon dwarf mistletoe (*Arceuthobium divaricatum*) is a parasitic plant that grows on pine trees and is colored in shades of yellow, orange, and red.

American cancer-root (*Conopholis americana*) is a parasitic plant with no green parts that grows on the roots of oak and beech trees.

58
What is the flattest plant you can find?

HINT: You may need to get on the ground for this one.

Plants usually grow up toward the light, sending shoots in multiple directions. However, if you find a nice moist ditch, a wet rock surface, or an exposed cutout of shady soil, you may find some of the flattest plants you'll ever see—and they are not each other's closest relatives.

Hornworts and liverworts are some of the flattest plants you will find. Along with mosses, they make up the three lineages of the bryophytes. Instead of growing upright like a moss, they grow by producing a flattened structure called a thallus. Fern **gametophytes** live a similar

lifestyle. Because these plants are so low to the ground and often shaded, they obtain sunlight through specialized light receptors, which can find the smallest flecks of light in the shadiest habitats.

They are also small because they lack vascular tissues and must reproduce with freely dispersing sperm that have to swim along thin films of water over several centimeters—which may as well be a double marathon for them. But these plants live and thrive. Being the largest plant out there is not the ultimate goal. These flat plants essentially live in two dimensions—and that is just fine for them.

You may find a hornwort eking out a living in moist ditches. Hornworts grow flat along the ground except for their "tall" reproductive structures, the horn-shaped sporophytes. A hornwort's primary body is no thicker than a few millimeters.

On a shady soil cutout you may find a tiny fern gametophyte, just one cell thick. This is the often-flattened, heart-shaped **haploid** phase of the fern life cycle.

On wet rock faces, you may observe a liverwort like this snakewort (*Conocephalum salebrosum*). Like the hornworts and unlike the mosses, many of the liverworts produce a flattened body that creeps along its substrate. However, they send up reproductive structures that tower a whopping several millimeters over their flat bodies.

This lady fern (*Athyrium angustum*) shows signs of abundant herbivory, likely from chewing sawfly larvae, which feed like caterpillars.

59

How many different signs of herbivory (plant eating) can you find?

Who has been munching on the plants around you? Many animals, from insects to birds to mammals, have evolved strategies to eat plants. Depending on their size, mouthparts, and life history, their eating process may leave behind very different traces. For instance, caterpillars are chewing insects, taking large chunks out of leaves. Some flies and moths are leaf miners, and their larvae burrow between the tissue layers of a leaf, eating the internal cells and leaving wandering tracks. Aphids and scale insects are sucking or piercing insects, sticking their strawlike mouthparts into the phloem. And large mammals chomp off large bits of the shoot system.

Each animal's mode of feeding is reflected in the visual damage. Take a look at a plant that's been partially eaten, and with a little practice there's a good chance you'll be able to identify the type of animal that was attacking it.

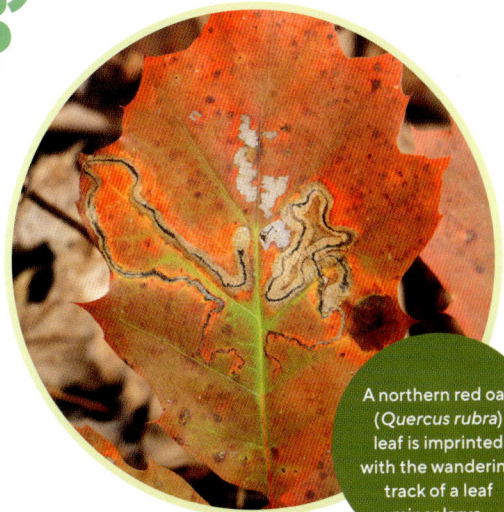

A northern red oak (*Quercus rubra*) leaf is imprinted with the wandering track of a leaf miner larva.

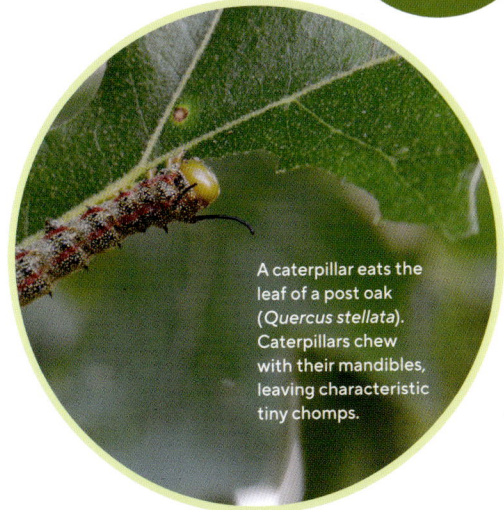

A caterpillar eats the leaf of a post oak (*Quercus stellata*). Caterpillars chew with their mandibles, leaving characteristic tiny chomps.

patterns

135

60
Can you find a plant that eats animals?

Animals eat plants. But plants have figured out a way to strike back. Though they are not often limited in sugar—they make it from scratch—they are limited in other nutrients like nitrogen. Some nitrogen can be obtained from special bacteria in the soil that take it from the atmosphere and turn it into a usable form for plants. However, some ecosystems, like bogs, significantly lack nitrogen. This means that many plants growing in bogs are nitrogen deficient. Bog-dwelling species like the round-leaf sundew (*Drosera rotundifolia*), purple pitcher plant (*Sarracenia purpurea*), and common bladderwort (*Utricularia vulgaris*) have evolved the ability to consume animal flesh to obtain their nitrogen.

Don't be too worried—this isn't *Little Shop of Horrors*. These plants are relatively small and tend to eat small invertebrates; however, every once in a while, a frog or salamander may fall prey. Each of these plant species has a distinct mechanism for capturing its prey. Once trapped, insects or small vertebrates are slowly digested with enzymes and sequestered by the plant—quite a macabre evolutionary innovation.

Interestingly, all of the carnivorous plants shown here are in the same clade, the Caryophyllales, even though they have highly diverse capturing mechanisms. But they are not the only ones. Other lineages have evolved carnivory in drastically different groups, including the tropical pitcher plants (*Nepenthes*) and even bromeliads. This suggests that carnivory has evolved multiple times independently across the flowering plant tree of life.

Leaves of round-leaf sundew (*Drosera rotundifolia*) are covered in numerous sticky trichomes where insects get caught. Digestive fluids will then be secreted from these sticky traps.

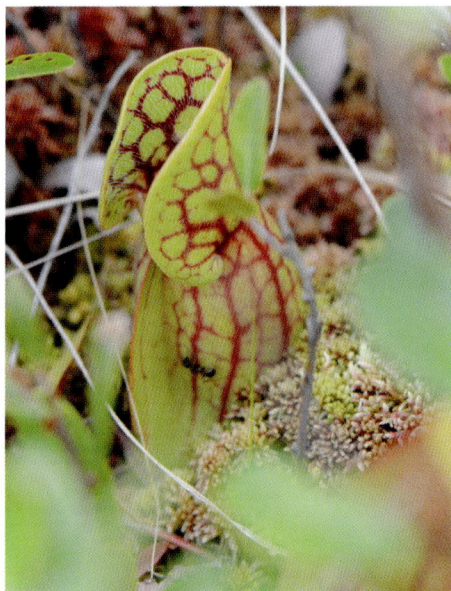

An ant crawls up the carnivorous pitcher of a purple pitcher plant (*Sarracenia purpurea*). When an animal falls into the pitcher, it gets stuck. The slippery surface has backward-facing hairs that allow them to fall in, but they can't crawl out.

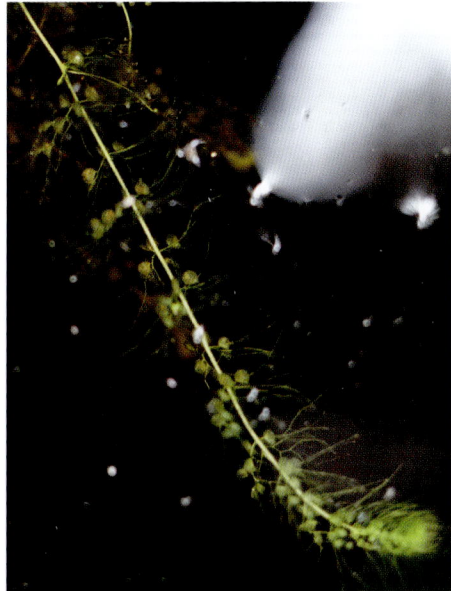

Bladderworts produce highly intricate vacuum structures that suck up their prey! The leaves of common bladderwort (*Utricularia vulgaris*) include small, globular traps that suck in and digest tiny aquatic invertebrates.

The leaves of an American horse chestnut (*Aesculus hippocastanum*) begin to unfurl from the bud.

61

Can you track a single leaf as it expands out of a bud?

Tulip poplar (*Liriodendron tulipifera*) leaves are visible through the protective stipules that surround the bud.

Nearly all the cells of a leaf are produced when the leaf is just beginning to grow. In many temperate trees, this happens in the preceding spring or summer, while the **leaf primordia** (the smallest beginnings of a leaf at the apical meristem) lie dormant in their resting buds. The following spring, these leaves will emerge from their buds and become orders of magnitude larger than they were mere days before. This is striking in plants like horse chestnuts, which expand their leaves to the size of dinner plates.

This growth happens almost exclusively by pumping the cells full of water, not by the division of cells into new ones. By the time a leaf is ready to expand from a bud, its cells have already been busy dividing. Many of the cells that will make up the mature leaf are already present but are only a fraction of their final size. To expand the leaf, the plant pumps water into the cells like a water balloon, and an average cell might expand 50 to 100 times in volume by the time it reaches its final size (see prompt 35).

62

Can you identify the oldest and youngest leaves on a branch?

In many animals, including ourselves, biological develop-ment happens in one phase of our lives and then stops. Here we are using the word *development* in the biology-of-organisms sense, meaning the growth of new organs, or **organogenesis**. Our own process of organogenesis hap-pens once in our lives, in utero, to produce the set of major organs that will grow in size and perhaps change in shape and condition, but they will never be added to or replaced by entirely new organs.

In contrast, plant development happens continuously over the course of a plant's life. Leaves are organs, roots

A black birch (*Betula lenta*) branch shows the sequence of leaves produced in one year.

are organs, and flowers are systems of organs. Organs are highly specialized collections of different cells and tissues that perform a specific function. In perennial plants, all these organs continue to be produced anew over time. What's more, they are produced in predictable places, which means we can always find the newest organs on a plant if we know where to look (see prompt 16).

New growth occurs at the tips of branches. Therefore, along a tree branch, the leaves closest to the trunk will be the oldest (if only by a matter of days), and the leaves at the tip, farthest from the trunk, will be the youngest.

Labrador tea (*Rhododendron groenlandicum*) shows older, larger leaves at the bottom of the branch, with newer leaves closer to the flowering end.

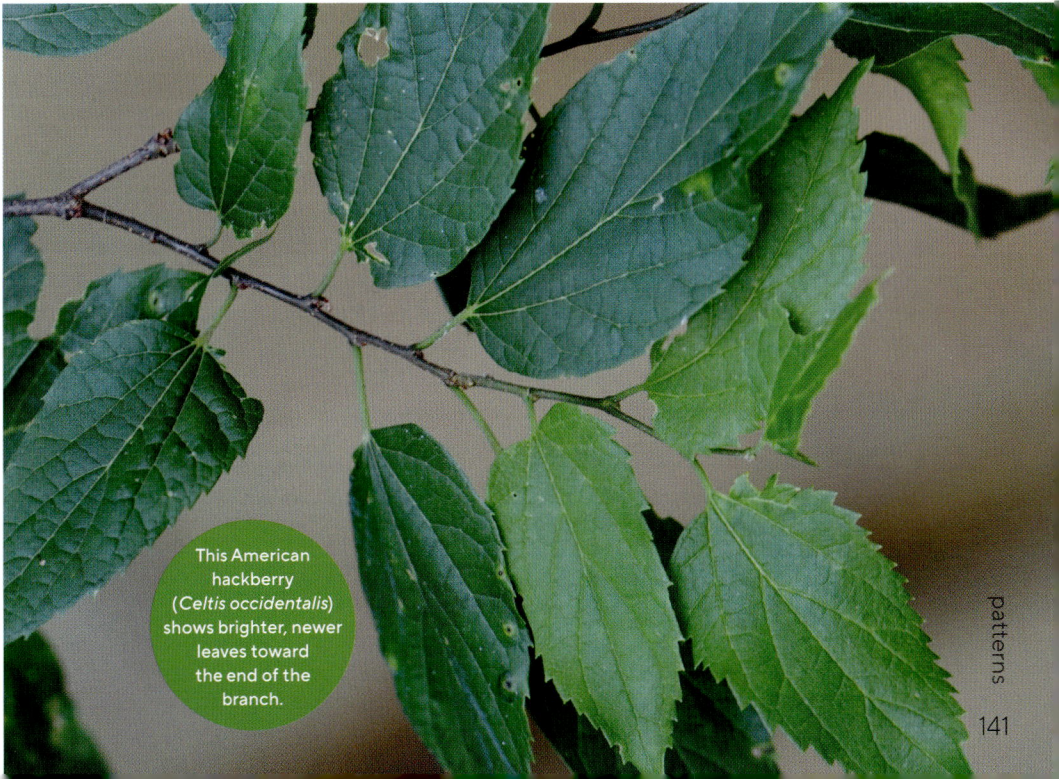

This American hackberry (*Celtis occidentalis*) shows brighter, newer leaves toward the end of the branch.

63

Arrange leaves from the same tree in a sequence of colors from green to brown.

HINT: Try this at the end of the growing season.

Healthy leaves are packed full of chlorophyll, but chlorophyll molecules don't last forever. Inside a leaf, individual molecules usually degrade after a few days, but they are constantly replenished. In temperate regions, when deciduous trees prepare to drop their leaves and enter a period of dormancy, the production of chlorophyll slows down. As less new chlorophyll is produced, pigment molecules called carotenoids contribute more and more to the overall color of a leaf (see prompt 46). Carotenoids are actually present in the leaf at all times of year, but their hues of yellow and orange are masked by green chlorophyll.

In many trees, color change stops here, and leaves will proceed from yellow or orange to brown as all types of pigment eventually break down. However, in some trees, as chlorophyll goes away and carotenoids are revealed, anthocyanins begin to be produced, adding shades of red and purple. Why would a plant invest resources in producing new pigments in a leaf that is preparing to be shed?

One theory is that anthocyanins provide protection from UV light, helping the nutrient reuptake process run more efficiently while the tree extracts as many nutrients as possible from its leaves before they drop. Another theory is that the bright anthocyanins may signal that a tree is capable of producing high concentrations of chemical defenses. This could dissuade herbivores like aphids from choosing that tree when they are seeking a host plant to spend the winter on. To date, neither of these hypotheses is fully supported, and they are not mutually exclusive.

Maple (*Acer*) leaves display a wide range of colors in the fall, including bright red.

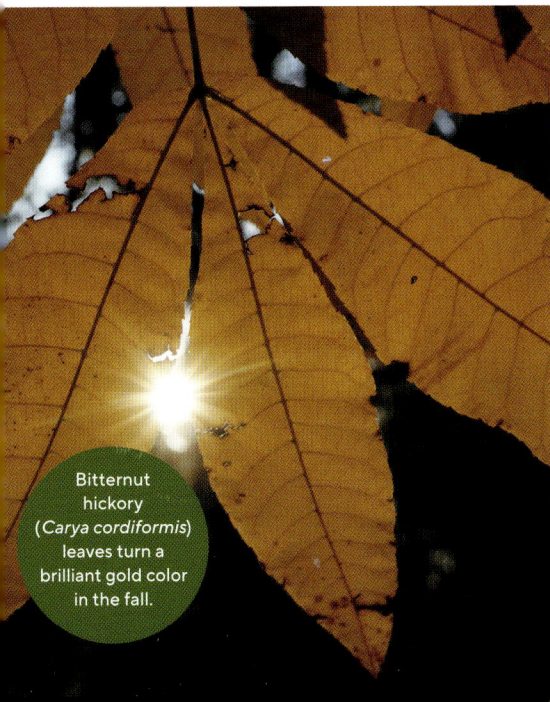

Bitternut hickory (*Carya cordiformis*) leaves turn a brilliant gold color in the fall.

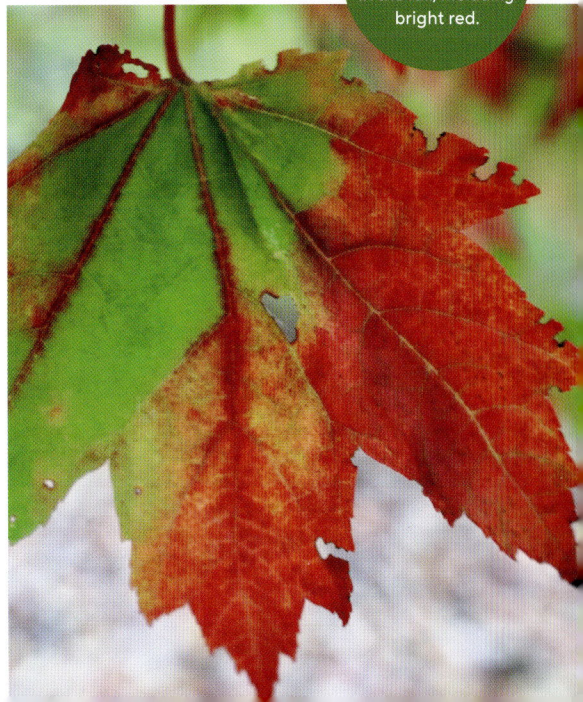

64
Can you find a winter tree with dead leaves still attached?

Most deciduous trees drop their leaves shortly after those leaves change color and die. However, in some species, dead leaves remain attached to the tree for a long time before finally dropping. We call this pattern **marcescence**. For example, a marcescent oak tree will retain a mantle of brown, crispy leaves on its branches throughout the late fall and winter and eventually drop them early the following spring before a new flush of leaves emerges.

The dead leaves clinging to a marcescent tree retain a small layer of living tissue at their base where they attach to the tree branch. This layer is called the abscission zone (see prompt 29). These living cells remain pliable and strong, which keeps the otherwise dead leaf firmly attached. The leaves can only fall off when this layer of cells finally dies.

Narrow-leaf spicebush (*Lindera angustifolia*) retains its brown, dead leaves.

A white oak (*Quercus alba*) holds on to dead leaves through the winter months, even covered in snow.

WHY HANG ON?

There are a couple of theories for why a marcescent habit might benefit a tree, and they are not mutually incompatible. First, the leaves on marcescent species remain photosynthetically active for longer than those of trees that drop their leaves more quickly. In cold climates, that means a tree can eke out every last opportunity to produce sugar before it becomes too cold for photosynthesis. Another theory is that dropping leaves in early spring could inhibit the germination and growth of potential competing plants around the base of the tree. Marcescent species also grow in tropical and high-alpine habitats, and scientists are still studying the potential evolutionary benefits in all of these places.

patterns

65
How many flowers of different colors can you find?

Flowers span the entire spectrum of visible colors— and even more in the UV spectrum. Many flowers are also boldly patterned in ultraviolet wavelengths that the human naked eye cannot perceive. Remarkably, the vast majority of this diversity in visible flower colors is created by just three classes of pigment molecules: carotenoids, anthocyanins, and **betalains**. Carotenoids produce yellow, orange, and reddish colors. They are the primary pigment in pumpkins, carrots, and sweet potatoes and are even responsible for the bright red color of cooked lobsters (via the marine algae that they consume) and the pink of flamingos (via the shrimp in their diet, which accumulate carotenoids from the algae they eat). Anthocyanins are responsible for reds, purples, and indigos. If they are concentrated enough, they can turn the petals or leaves black. These are the pigments underlying the color of red roses, blueberries, and purple violets.

Betalains, which also create pinks, purples, and reds, are a unique class of pigment molecules that are only produced by plants in the order Caryophyllales. One familiar example is the deep magenta color of beets (*Beta vulgaris*), from which betalains were first described and for which they are named. This order also contains the cacti, many of which produce bright, betalain-colored fruits.

A yellow violet (*Viola*) is a great example of a carotenoid-colored flower.

Smooth phlox (*Phlox glaberrima*) produces bright pink flowers colored predominantly by anthocyanins.

Great bougainvillea (*Bougainvillea spectabilis*) flowers are surrounded by bright pink bracts, which are colored by betalain pigments.

Forsythia (*Forsythia*)

California poppy (*Eschscholzia*)

Dahlia (*Dahlia*)

Coneflower (*Echinacea*)

Lilac (*Syringa*)

Monkeyflower (*Mimulus*)

Galls on Utah juniper (*Juniperus osteosperma*) are produced by a small fly in the genus *Walshomyia*.

66
Can you find an unusual growth on a leaf?

Witch-hazel cone galls grow on American witch hazel (*Hamamelis virginiana*) and are produced by an aphid.

An oak apple gall is caused by an oak gall wasp.

Leaves usually have flat surfaces, so when something that looks like a wart, bubble, or tumor appears, it stands out. These abnormal outgrowths are **galls**, and they can be caused by several different groups of arthropods (including mites, flies, and wasps), fungi, bacteria, viruses, or nematodes. That's a long and diverse list of potential gall-forming agents, but they all share the same general strategy. In each case, a gall is initiated when the agent either lays an egg on or infects the plant. The egg or infection then manipulates the surrounding plant tissue to grow in ways it normally wouldn't.

Every plant structure, from leaf to stem to flower, develops according to a set of instructions encoded in the plant's DNA, with additional influence from the environment. Gall-forming agents chemically upregulate a plant's growth hormones, which kick-starts the development of the gall. This can be thought of as a tumor, but one caused by temporary external manipulations of hormonal signals rather than mutations in the plant's DNA (like a cancerous tumor).

But how do gall-formers benefit by manipulating plants in this way? By inducing the growth of a gall, these species coerce the host plant into building a home for them to shelter the eggs or colonies of their next generation.

67
Can you find a plant growing on a rock?

Boulder broom moss (*Dicranum fulvum*) is typically found growing on rocks; this species, like all mosses, lacks roots and holds on with rootlike projections called rhizoids.

Plants have figured out a way to live in nearly every kind of habitat. Some plants grow on the forest floor, some in the water, and some on other plants. Certain species have even found a way to eke out a living on rocks. These radical plants are called **epipetric**, from the Greek for "upon rocks." This lifestyle is not easy. With essentially no soil to hold stores of water, epipetric species need to deal with drought, and their roots must pave their own way, grabbing hold for dear life.

The chemistry of different rocks can also impact these species. For instance, limestone has an extremely high pH of nearly 10 (very basic), while serpentine rocks have very low levels of many essential nutrients (such as nitrogen) and high concentrations of heavy metals like chromium, cobalt, and nickel. Interestingly, some species love these harsh, rocky substrates. The maidenhair spleenwort (*Asplenium quadrivalens*) is a lover of limestone, and the New England endemic Green Mountain maidenhair fern (*Adiantum viridimontanum*) is a resident of serpentine outcrops.

Many evolutionarily distant lineages like mosses, lycophytes, ferns, and flowering plants can be epipetric. While rocks are not the richest habitat, they are a habitat nonetheless.

This pineweed (*Hypericum gentianoides*) is growing from a crack in a large boulder.

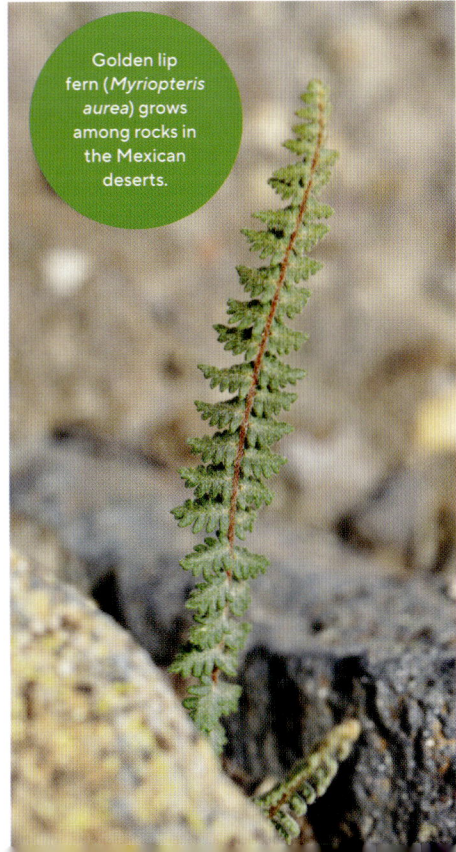

Golden lip fern (*Myriopteris aurea*) grows among rocks in the Mexican deserts.

68

Can you find a plant that produces pollen and seeds in separate flowers?

HINT: These flowers might be on different parts of the same branch.

Do you see the two kinds of inflorescences on this birch? The long, pendulous ones produce the pollen, and the bright green ones sticking straight up produce the ovules.

Some flowers are poetically called perfect. This is not a value judgment but a term referring to flowers that contain both pollen-producing parts (stamens) and ovule-producing parts (carpels). **Perfect flowers** produce all components necessary for reproduction, but successful sexual reproduction does not *require* this "perfect" state of affairs.

Indeed, many flowering plants produce "imperfect" flowers, which are missing either stamens or carpels. In some cases, both types of imperfect flower will be present on the same individual; this is a **monoecious** plant. Plants such as the birches are monoecious. In other species, an individual plant will bear only pollen-producing or only ovule-producing flowers, but not both; these plants are **dioecious**. For instance, many holly plants in the genus *Ilex* are dioecious, and only the ovule-producing individuals will bear the characteristic red berries. In both monoecious and dioecious plants, the production of pollen and ovules in different flowers reduces the chances that a plant will mate with itself.

While some plants primarily or exclusively self-pollinate, other species have evolved additional strategies to avoid inbreeding. Being able to mate with yourself provides insurance for reproductive success and the production of another generation of seeds, but inbreeding comes with the risk of exposing the next generation to harmful genetic variants. Many species of flowering plants have chemical systems in place to ensure that they will not accept their own pollen but will allow pollen from other individuals of their species to fertilize their ovules.

TOP: Only ovule-producing holly plants (*Ilex*) will produce red berries—and must be pollinated by a pollen-producing individual in order to do so.

MIDDLE: These hanging clusters of American hornbeam (*Carpinus caroliniana*) catkins are its pollen-producing inflorescences. This species is monoecious and produces drastically different pollen- and ovule-producing flowers.

BOTTOM: These ovule-producing flowers of American hazelnut (*Corylus americana*) show their red stigmas. This species is also monoecious.

patterns

69
How many different plants can you find that freely disperse spores?

Some ferns produce entire leaves or leaflets devoted to spore production. An unfurling frond of royal fern (*Osmunda spectabilis*) shows the fertile, spore-producing section of the leaf covered in small, green, beadlike sporangia that house the spores.

In the lycophytes, sporangia are always produced in the upper side of the leaf axil—the point of attachment to the stem. Groundcedar (*Diphasiastrum digitatum*) is a lycophyte that produces spores in upright, fingerlike cones called **strobili**.

Not all plants produce flowers, fruits, or seeds. In fact, the first land plants—and many that are still around today—reproduce via freely dispersed spores. Ferns (see prompt 32), **lycophytes**, and mosses reproduce by propelling individual spores into the wind and hoping they land somewhere hospitable (see prompt 31). This trait did not evolve individually within each of these groups, but it is the original state of all the land plants— otherwise known as the embryophytes.

Spores are products of meiosis, meaning they contain half of the genetic material of the parent plant. They are produced in spore houses called sporangia.

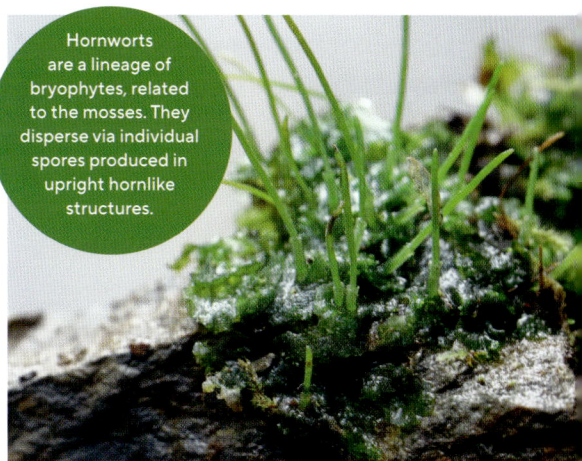

Hornworts are a lineage of bryophytes, related to the mosses. They disperse via individual spores produced in upright hornlike structures.

SPORE DIVERSITY

Spore sizes and shapes vary drastically, making them helpful in identifying different species. Spores can be spherical, oblong, or triangular. They can be smooth or have spinelike projections. And they are often very small, between 10 and 1,000 micrometers (around 500 times smaller than a rice grain).

Mosses are a bit different. The leafy, green part of the moss plant we commonly observe is haploid— meaning it has one set of chromosomes—and will produce eggs and sperm in small structures on its body. Once fertilization occurs, it will produce a small diploid organism—meaning it has two sets of chromosomes—called the sporophyte. The sporophyte will grow tall and eventually produce spores within a sporangium sitting atop a large stalk.

70
Compare a ripe fruit from your kitchen to one you find outside.

In kitchens and grocery stores, the term *fruit* refers specifically to sweet, fleshy fruits. These are typically fruits that we expect to eat without cooking, once we perceive them to be ripe. As we grow up, we quickly develop an intuition for what ripeness looks, feels, smells, and tastes like in these fruits of commerce. Usually, it means the color becomes less green and more intense, and the fruit may become softer to the touch and smell or taste sweeter.

In the wild, fleshy fruits have often evolved to be dispersed by animals, and it benefits them to be eaten so that their seeds can spread. However, plants need to spread mature, developed seeds; immature seeds cannot yet germinate and grow. As the seeds mature, these once-unpalatable fruits ripen into a more enticing form that is ready to be eaten. Fleshy fruits often change from green to more vivid colors, like yellows or reds, that contrast with foliage as they ripen, catching the eye of would-be fruit eaters. These animals are often mammals or birds, big enough to consume the fruit without destroying the seeds (see prompt 53).

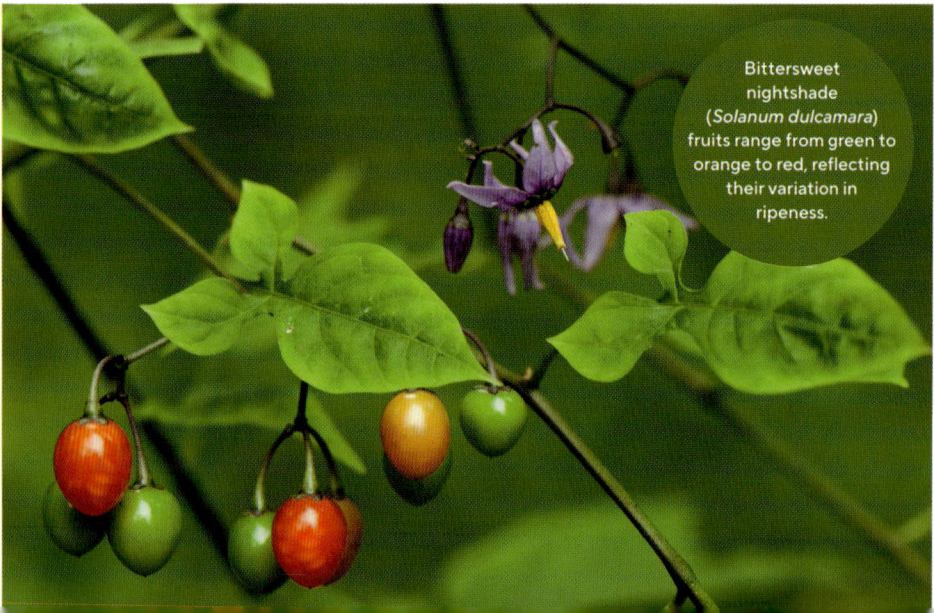

Bittersweet nightshade (*Solanum dulcamara*) fruits range from green to orange to red, reflecting their variation in ripeness.

A ripe persimmon is soft, sweet, and brightly colored.

READY TO TRAVEL

All fruits, whether they are fleshy or not, have a maturation process that increases the likelihood that their seeds are spread when they are fully formed. In non-fleshy fruits, this might mean that an outer casing cracks open to reveal or release the seeds (as with milkweed) or that a structure grows to catch a passing breeze (as with dandelions). While fleshy fruits start hard and soften as they mature, many other fruits, like those of maple and beech, start soft before maturing into hard, dry forms that can disperse in the wind or by other means.

Unripe, green fruits of American persimmon (*Diospyros virginiana*) do not yet have mature seeds and are not palatable to eat.

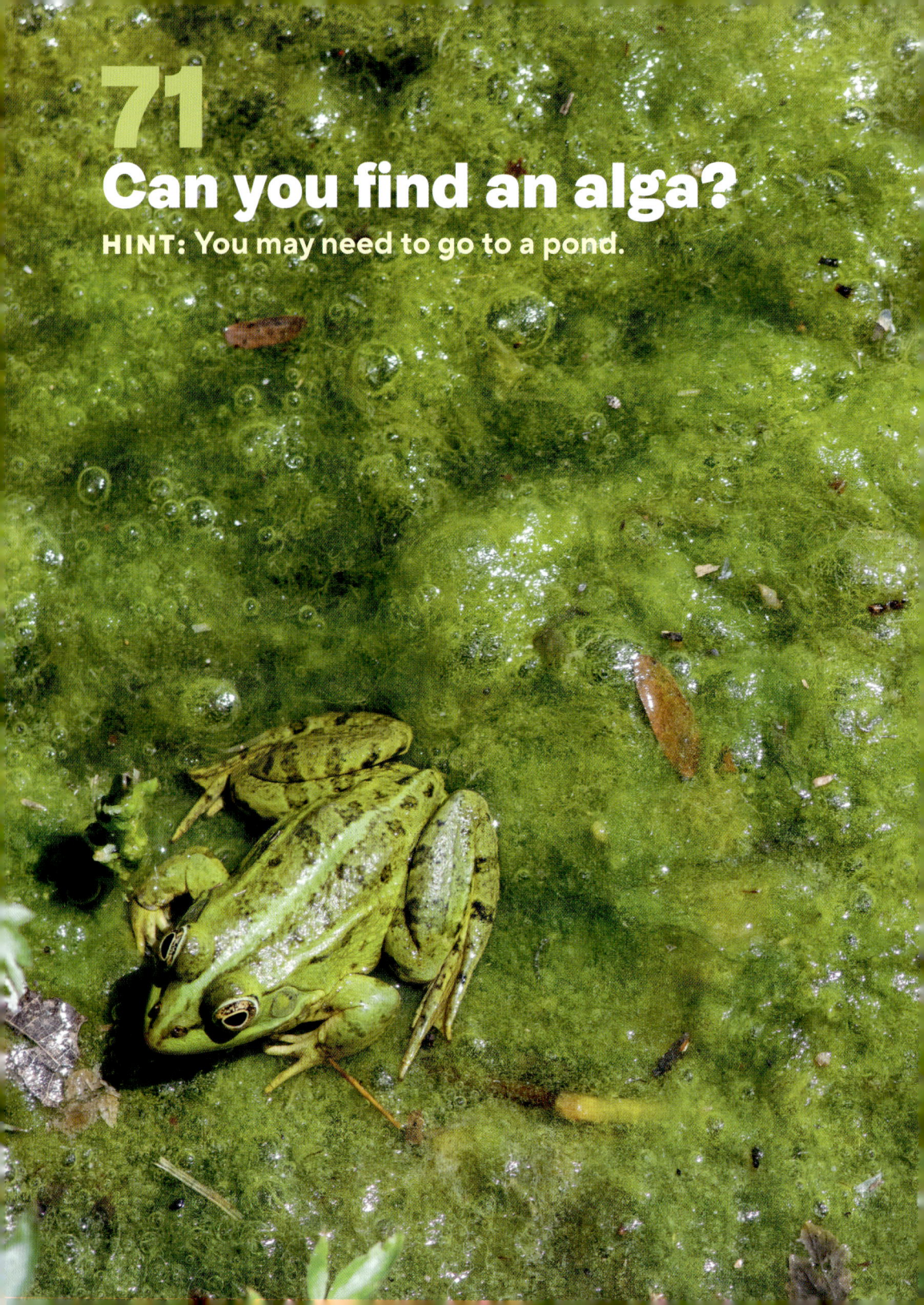

71

Can you find an alga?

HINT: You may need to go to a pond.

If you have ever been wading through a mucky pond, you may have noticed some green slime on your boots. This "slime" is not inert goop but living, breathing organisms. Commonly, these are members of a freshwater group of algae called **charophytes**, and, yes, they are plants (though some organisms we call alga, like brown and red algae, are not closely related to plants and have independently evolved the algal habit).

The way we define a plant has changed drastically over time. Over a century ago, they were defined as any **sessile** organism (meaning fixed in one place) that wasn't an animal (like a barnacle or coral). This included organisms like fungi, which we now know are more closely related to animals than plants. However, these days we take a more phylogenetic, or evolutionary, perspective. All plants are members of the **Viridiplantae**, a group that includes marine and freshwater algae and all of their descendants, including the land plants.

Back to the pond scum—a rough name we sometimes give to these foundational plants. Members of the charophytes include the genera *Chara*, *Coleochaete*, and *Zygnema*. From these groups of plants and their relatives, the modern land plants were born. Charophytes are freshwater aquatic algae that have evolved a simplified body plan with some, but not all, of the traits of land plants. Next time you are out in the pond, take a moment to appreciate the algae.

Not all organisms that we call algae are members of the charophytes. This unassuming brown alga grows in a freshwater pond.

An alga (cyanobacterial film) grows alongside aquatic plants in a greenhouse.

patterns

Pando spreads over more than 100 acres and weighs an estimated 6,000 metric tons, making it the heaviest known organism in the world.

72
Can you find a large patch of one plant species?

Are they separate individuals, or do they seem to be connected underground?

While walking in the woods, you may come across a dense expanse of an understory plant like hay-scented fern (*Sitobolium punctilobulum*). Although most plants reproduce sexually, many are also capable of (asexual) **clonal spreading**. In this case, what appear to be many distinct individual fern plants are all connected by an underground stem system. When plants spread clonally, they don't invest resources into growing one individual stem to massive size but rather in growing many interconnected units of modest size.

Clonal growth isn't limited to herbaceous plants. Some large tree species

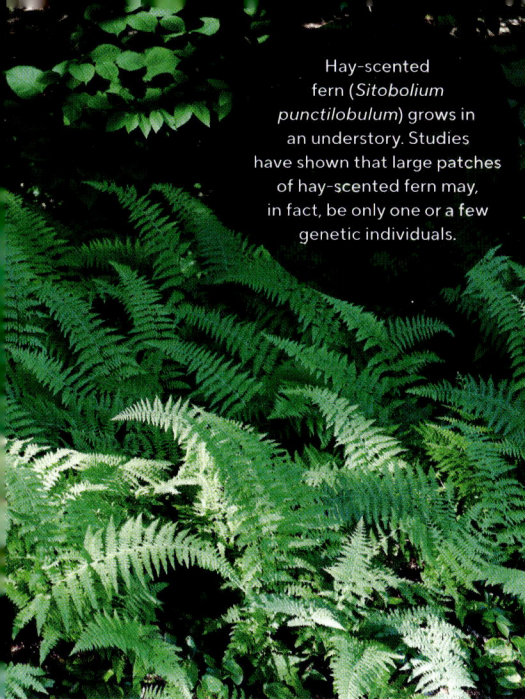
Hay-scented fern (*Sitobolium punctilobulum*) grows in an understory. Studies have shown that large patches of hay-scented fern may, in fact, be only one or a few genetic individuals.

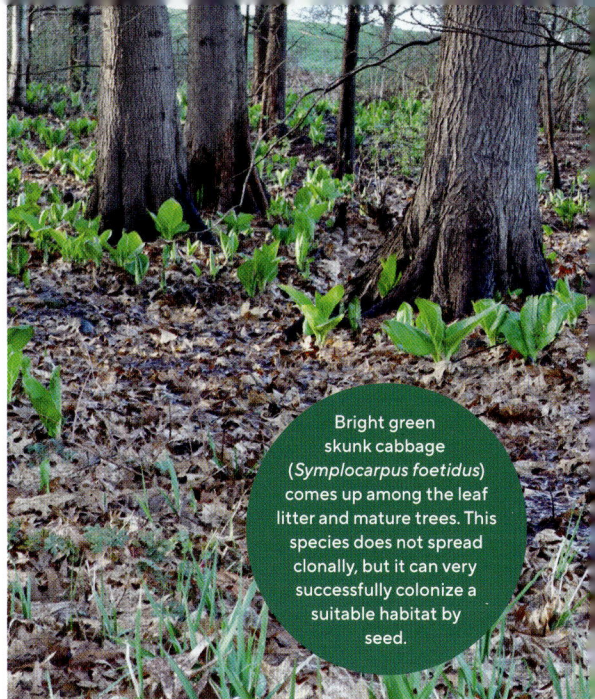
Bright green skunk cabbage (*Symplocarpus foetidus*) comes up among the leaf litter and mature trees. This species does not spread clonally, but it can very successfully colonize a suitable habitat by seed.

spread clonally, including quaking aspen (*Populus tremuloides*). One quaking aspen living in Utah has been named Pando. Pando looks like a grove of over 45,000 quaking aspen trees, but genetic tests have shown that all of these stems are connected belowground and actually represent a single individual organism.

SUPERLATIVES

Another clonally spreading plant, a colony of the aquatic flowering plant Neptune grass (*Posidonia oceanica*) in the Mediterranean Sea, is thought to be the oldest living organism at an estimated 100,000 years old! Clonal plants test the limits of what is possible for an individual organism.

patterns

161

A red maple (*Acer rubrum*) shows barbed-wire scars in its bark at the height of a former fence.

73

Revisit a tree with carving in its bark. Has the location of that carving ever changed?

Perhaps you were childhood sweethearts. You walked up to the smooth-barked beech tree and carved your initials together enclosed within a heart. One, two, three years later, your love has grown and so has the tree, but the heart-enshrined initials are stable where you carved them. Decades pass, and the tree is nearly 100 feet tall, but your grandkids can still see the evidence of the love you shared at the same height on the trunk. Why has the carving stayed exactly in place, even though the tree has continued to grow all this time?

This is because plants grow in height from the tips of their bodies, not from the base of their stems. When the carved beech tree was young, it likely had one growing tip. This tip is called the shoot apical meristem (SAM) . The SAM (see prompt 3) is found at the terminus of each branch. There are a set of cells in this region that are microscopic and **meristematic**, meaning they contain stem cells that can develop into different plant organs. As they expand and divide, they push the plant tip up toward the light.

As the cells develop below the SAM, they are locked in rigid place by their **cell walls**, meaning they can never, and will never, move relative to one another. This way of growing means that any given part of a trunk will never move from its current position. A carving will remain forever at the same height. While trees are resilient beings, carving into their bodies with a knife can expose them to infection—it's always best to leave their bark alone.

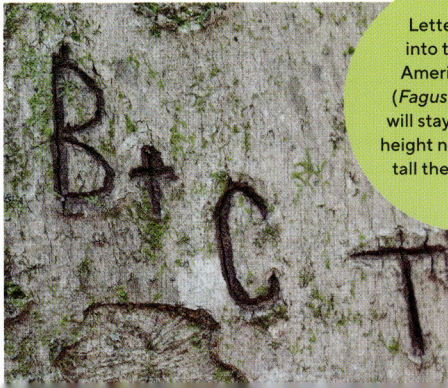

Letters carved into the bark of American beech (*Fagus grandifolia*) will stay at the same height no matter how tall the tree grows.

74

How many different flowers can you find that have five petals?

Caucasian pear
(*Pyrus caucasia*)

Petal number is an oddly specific trait, but it is quite diagnostic of large groups of plants. For instance, the monocots—plants including the lilies, grasses, palms, and their relatives—have petals in whorls of three (most often there are three or six petals). On the other hand, the **Magnoliids** have several to dozens of petals or petal-like structures.

But the most diverse and species-rich group of flowering plants are the **Pentapetalae**, or the five-petaled plants. These include roses, maples, beans, mints, daisies, and their relatives. There are approximately 300,000 species of flowering plants in this group, comprising almost 65 percent of angiosperm species richness.

At this point, you may be thinking of the last rose you saw. That had way more than five petals. That is because it was bred by horticulturalists to have many more petals than a naturally occurring species (see prompt 97). What about the mints or beans? If you take a closer look, you will notice that even in plants with tubular flowers, there are small notches on the tips as well as grooves lining the flower—clues that some or all of the five petals are connate (fused together).

WHY FIVE?

Is there something special about having five petals? Well, not exactly. Not every observed trait is an adaptation. Sometimes a trait just works, and evolution progresses without much change to that trait. In this case, it is likely that the common ancestor of the Pentapetalae had five petals and reproduced just fine, so over evolutionary time, selection acted on other aspects of floral morphology like shape, structure, fusion, and reduction, without changing the number of petals.

patterns

Phlox (*Phlox*)

Geranium (*Geranium*)

Periwinkle (*Catharanthus*)

Hibiscus (*Hibiscus*)

Columbine (*Aquilegia*)

Forget-me-not (*Myosotis*)

The common name *spicebush*, as with this Japanese spicebush (*Lindera obtusiloba*), refers to the sweet and spicy odor of the leaves when crushed.

75
Crush up a leaf and describe its smell.

Plants are complex pharmaceutical factories.

Whether it is the mint in our tea, the black pepper in our pasta, or the basil on our pizza, we often interact with the diversity of tastes and scents that plants can make. Plant smells, and many tastes, come from complex chemicals called **volatile organic compounds (VOCs)**. Plants do not produce these chemicals for human benefit but for their own protection. In many cases, VOCs are pesticides used by the plants to protect

Mustards (in the Brassicaceae family) produce sulfur-containing biochemicals called glucosinolates. These smelly compounds are hypothesized to have evolved for defense.

Raceme catmint (*Nepeta racemosa*) has strongly aromatic leaves, like other members of the mint family (Lamiaceae).

themselves against herbivores or other pathogens that may want to eat or attack them (see prompts 59 and 66).

Oddly enough, these chemicals can be just as toxic to the plants as they are to the organisms they are fighting. To ensure an individual plant doesn't poison itself, many species store these chemicals in secure balloon-like structures within their cells called **vacuoles**. These will burst upon damage from an herbivore, releasing the VOCs and hopefully deterring the threat.

It is no coincidence that plants are capable of producing a much greater diversity of chemical compounds than animals. Since plants can't move, they depend on chemical defenses (and mechanical ones, such as spines) to avoid being eaten. A similar explanation applies to fungi, which also produce fantastically diverse chemicals. Specific VOCs can be great taxonomic indicators of large plant groups like the mint family (Lamiaceae) or the mustard family (Brassicaceae).

patterns

76
How many different floral scents can you find?

Plants are some of the only organisms that we intentionally smell. It is probably fair to say that botanizing is the most smell-engaged natural history hobby. Flowers in particular are often celebrated for their enchanting fragrance. In fact, some flowers are appreciated more for their smell than for how they look (although they usually look great, too!). Lilac (*Syringa*), peonies (*Paeonia*), and milkweed (*Asclepias*) all have flowers renowned for their scents. These species and many more are commercially important in the perfume industry, reflecting the value humans place on floral scents.

Of course, these plants did not evolve their enticing odors for the benefit of humans. The smell of a flower is a signal broadcast into the air to attract other animals that might provide pollination services. This comes in especially handy at nighttime, which is why flowers that have evolved to attract night-flying pollinators like bats and hawkmoths produce strong scents. Biochemically speaking, floral odors, like leaf odors, are due to volatile organic compounds, which are typically gases at room temperature.

The sweet-smelling scent compounds produced by peony flowers (*Paeonia*) are extracted as a perfume ingredient.

Common milkweed (*Aslecpias syriaca*) flowers produce a pleasant, sweet aroma that is sometimes compared to vanilla.

77

Hunt for lichens and compare their textures.

HINT: Look on rocks, trees, and even the forest floor.

Lichens were initially thought to be plants, but they are actually far from them. A lichen is a symbiosis between a fungus and either an alga or a photosynthetic bacterium called cyanobacteria. Sometimes there are also fungal yeast thrown into the mix.

Lichens grow on trees, rocks, and even in marine environments. The fungus, which is often a cup-forming fungus (ascomycete), forms the structure that provides the basic body of the lichen. The algae or cyanobacteria live in specialized structures that the fungus makes and produce sugars by photosynthesizing. The algae or bacteria receive a home, and the fungus receives a meal—a classic example of symbiosis.

You might think that such an intricate relationship would create dozens, hundreds, thousands, or even millions of distinct individual symbiotic relationships, no single one similar to another. However, lichens do form distinct "species" or "types." This occurs in part because of the way they reproduce. Many lichens can reproduce asexually, where small chunks of their bodies break off and disperse to new environments. These pieces, or **propagules**, include both the fungus and the alga or bacterium and have the potential to form whole new individuals.

The elaborate textures, colors, and even biochemistry of lichen allow them to inhabit nearly all corners of the earth, including some of the harshest terrestrial conditions on the planet. They are quite common on the Arctic tundra, where most plants struggle to survive—in these ecosystems, lichen comprise a significant portion of aboveground biomass.

The texture of a lichen is mostly dictated by the fungal symbiont. Some produce branchlike structures (called fruticose), while others are crustose, forming dense, rough mats. Reindeer lichen (*Cladonia rangiferina*) is a fruticose species, growing in a bristly carpet on the forest floor.

In moist habitats, tree branches are often bursting with fantastical lichens like this beard lichen (*Usnea strigosa*).

Common greenshield lichen (*Flavoparmelia caperata*) grows in papery flakes, splattered along a tree trunk.

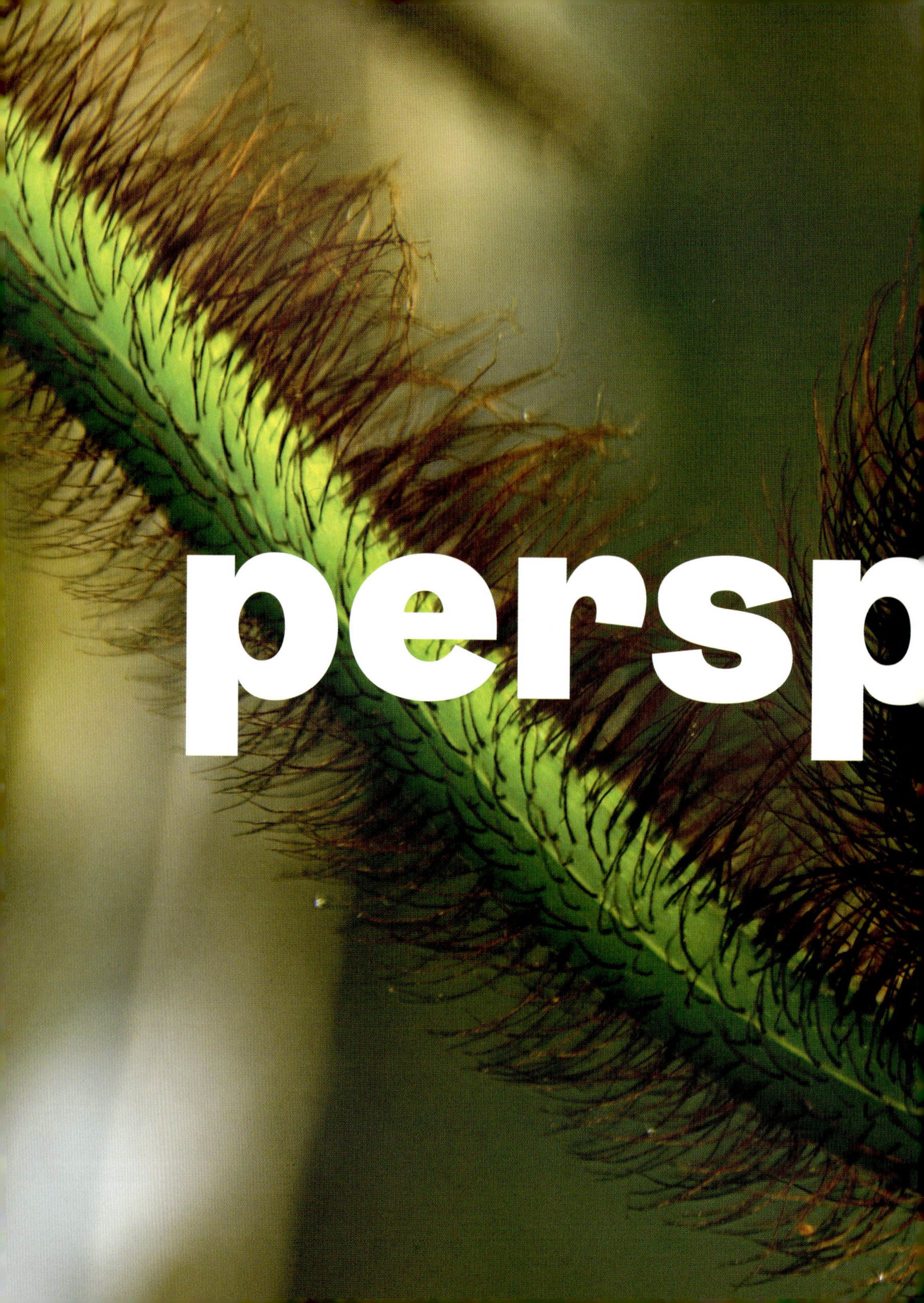
persp

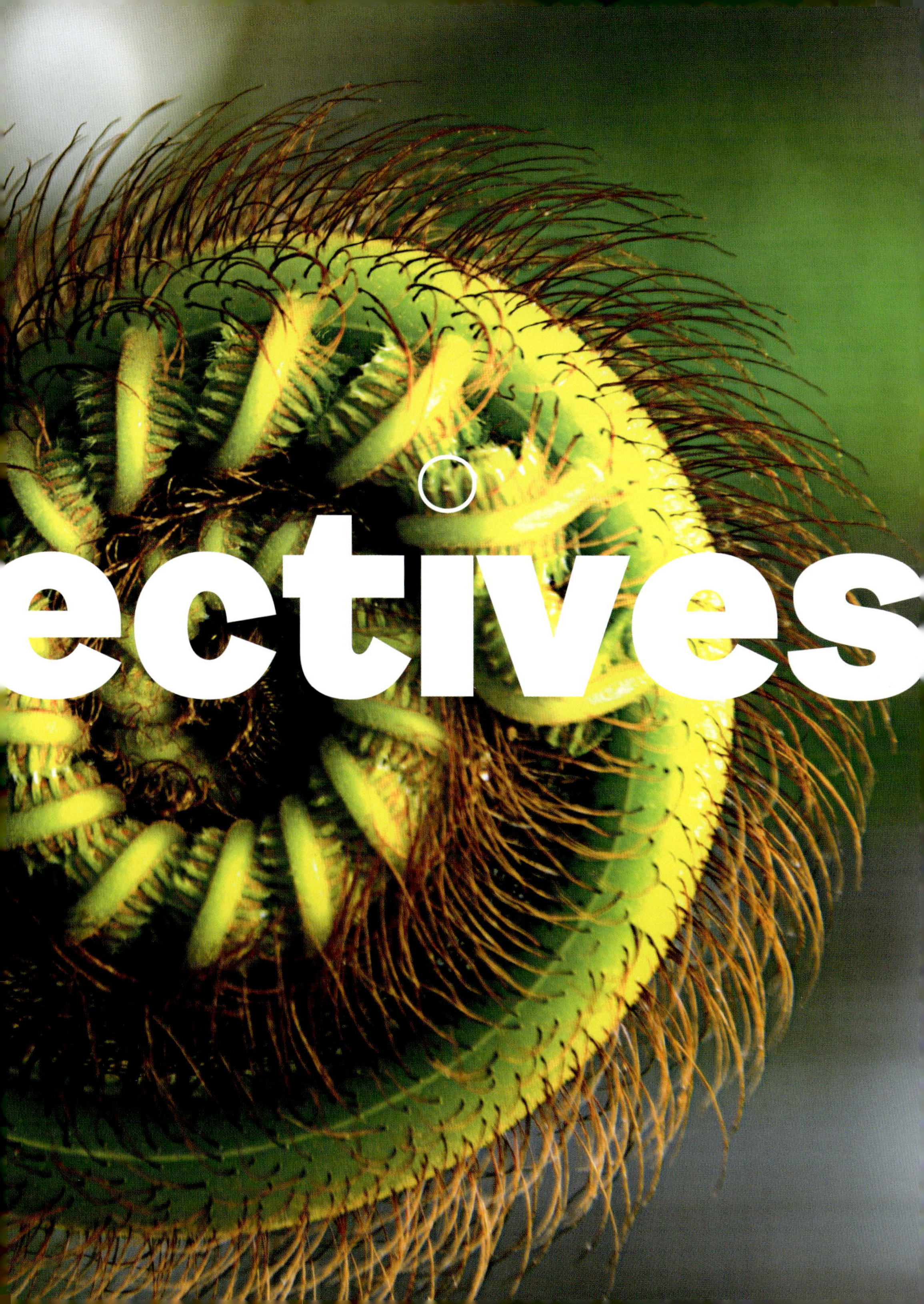
ectives

78
What was the first plant you noticed today?

It shouldn't take long after waking up to notice your first plant of the day.
Plant materials are so thoroughly embedded in our lives that they are inescapable. Our pillowcase or clothing (cotton, linen, and rayon), our furniture, the structure of our homes, and much of the food we eat may all be made from plants.

We are deeply dependent on the plant kingdom in every aspect of our daily lives, and this is a testament to the chemistry that plants can perform. Beginning with the process of photosynthesis, plants turn carbon dioxide into various compounds that may be flexible and turned into thread, or sturdy and turned into building materials, or edible and turned into a meal.

If you try to notice the first living plant in your day, in nearly every location on Earth, you will quickly find plants and plenty of them. Just look out your window, or take a step outside. Only extreme environments like the North and South Poles are not dominated by plants.

Morning sunlight pierces through a branch of bitternut hickory (*Carya cordiformis*).

Coffee "beans" (*Coffea arabica*) do not actually belong to the bean family (Fabaceae), but rather the Rubiaceae family. They may be among the first plant parts you encounter in your day.

79
Keep track of the number of plant species you eat today.

Your morning oatmeal, biscuit, and cup of tea or coffee are all made from plants. So are the bread, lettuce, and tomato from your lunch sandwich and the pasta in your dinner. Even the vanilla ice cream you have for dessert is flavored by an orchid. These are all examples of plants we eat each day. But if you move up the food chain, the list expands exponentially. The chickens that laid the eggs you ate for breakfast consumed plant seeds, and the cow of your steak dinner consumed grass. At one point or another, nearly all energy used by both carnivores and herbivores passed through plants.

The only way energy is put into Earth's biosphere is through photosynthesis. Plants are the primary producers of nearly every ecosystem. This means that they convert the energy of the sun into usable

Wheat (*Triticum*)

biological energy for other organisms. Indeed, plants must be eaten for all other life on Earth to survive.

But plants clearly don't want their bodies to be eaten. This creates an interesting situation where plants continue to evolve anti-herbivory strategies, but paradoxically, if they were to become too good at evading predation, they would destroy the intricate web of biodiversity that they rely on. This has never happened in evolutionary history, nor is it ever likely to happen, but it's a reminder of the delicate balance of life.

TOP: Cranberries (*Vaccinium macrocarpon*) growing wild in a bog are the same species we cultivate and eat.

MIDDLE: Pyramids of spices fill the bins in a market in Marrakesh, Morocco. Black pepper, cinnamon, paprika, cardamom, thyme, sage—all herbs and spices come from living, growing plants.

BOTTOM: The oats in your breakfast oatmeal came from the grass *Avena sativa*. You may top that off with blueberries (*Vaccinium corymbosum*). If you like pancakes in the morning, they're probably made from wheat flour (*Triticum aestivum*) with some syrup from sugar maples (*Acer saccharum*).

80

During your next meal, pay attention to how many different parts of a plant you eat.

Not only do we eat a diversity of plants, but we eat a diversity of plant organs. We eat the leaves of lettuce, the petioles of celery, the bark of cinnamon, and even the sap of maples (see prompt 79). From a plant's perspective, it is generally worse for an animal to eat its reproductive structures than other parts like leaves or bark. However, in some cases, as with capers (*Capparis spinosa*) and broccoli (*Brassica oleracea*), we do eat small unopened floral buds. With saffron (*Crocus sativus*)—the most expensive spice at roughly $10 to $20 per gram—we consume the stamens, the structures that produce pollen.

Tea and herb market in Marrakesh, Morocco

When we eat blueberries, tomatoes, oats, corn, and flour, we are eating the fruits or seeds. Some plants have evolved fruits that attract hungry animals and take advantage of animal mobility to disperse their seeds. In such cases, the seed itself is capable of passing through the gut of an animal unscathed or may even require such a passage to become primed for germination. Most fleshy fruits, such as blueberries and raspberries, use this strategy. However, when an animal eats and breaks down the seed itself, this does not aid in dispersal but rather ends the possibility of successful reproduction for that individual seed. Indeed, it is important to distinguish between herbivory, which often does not kill the whole plant, and seed predation, which kills an embryonic plant.

Nonetheless, humans and many other animals love to eat the seeds of grasses (such as rice, wheat, and corn), trees (such as walnuts, almonds, and cashews), legumes, and other plants. Next time you sit down to eat, take a moment to reflect on the number of species and diversity of plant parts filling your plate.

Walnut (*Juglans regia*)

Strawberry (*Fragaria*)

Celery (*Apium graveolens*)

Saffron (*Crocus sativus*)

Cinnamon (*Cinnamomum verum*)

Caper (*Capparis spinosa*)

The old and scraggly architecture of a Utah juniper (*Juniperus osteosperma*) shows how it has moved in response to its environment over centuries.

81
Spend 10 consecutive minutes with one tree.

What do you imagine it is doing?

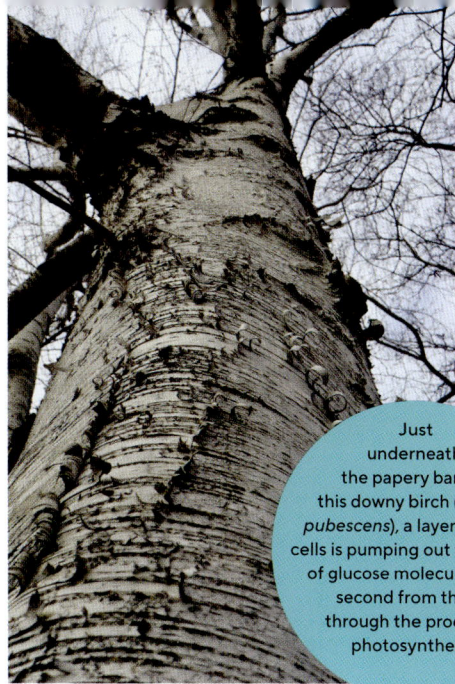

Just underneath the papery bark of this downy birch (*Betula pubescens*), a layer of green cells is pumping out thousands of glucose molecules every second from thin air through the process of photosynthesis.

A tree performs all the necessary functions of life while standing perfectly still. Or at least that is how trees appear to us. In fact, they are highly active, just on different scales of space and time than humans are. Humans and most other animals physically manipulate the world to perform the basic functions of life. We chew and swallow food to acquire energy, we expand and contract our lungs to exchange oxygen and carbon dioxide, and we walk to move about our environment.

Rather than manipulate the world with large-scale movements, plants perform their primary activities at the microscopic level of chemistry and physics. When we need to acquire water, we move our bodies to a water source, gather it with our hands, and then swallow the water through our mouths. In contrast, the body of a plant is built with an intricate network of xylem conduits that take advantage of the physical properties of water to pull large amounts from the soil without moving an inch. A mature tree will move tens to hundreds of gallons of water per day through its body. Oxygen and carbon dioxide flow in and out of leaves and stems through stomata and **lenticels** (see prompt 100) without any structures like lungs. Meanwhile, a leaf may look still, but inside of its cells, molecular machinery is buzzing along, hurling electrons from one molecule to another and performing many chemical reactions per second to create sugar from carbon dioxide and water.

On longer timescales, trees detect environmental cues about day length and temperature that guide their annual cycles of growth and dormancy. Most of these processes are too small, too fast, or too slow for us to easily perceive with our unaided senses, and yet both trees and humans work each day to meet the same basic requirements for living.

perspectives

82
How old is the wood in your furniture?

HINT: Look closely at the wood grain.

Are you sitting on a tree right now? Or maybe this book was on a tree before you picked it up to read some prompts. Whatever your interior design decisions are, there is certainly a piece of wood somewhere in your home. Take a look at it. What do you see? You should see waves, ridges, lines, and holes. These are the complex patterns of cells making up the tissue called xylem, the main component of wood.

The xylem is the main site of water movement and carbon storage in a trunk. It provides the structure and integrity of the whole plant, and this is why we use it as a building material. The structure of the cells and their associated functions are what give your wood furniture its style. The holes you may see are water-conducting cells, **tracheary elements**. These are wide, hollow tubes that move water through the tree. The distribution

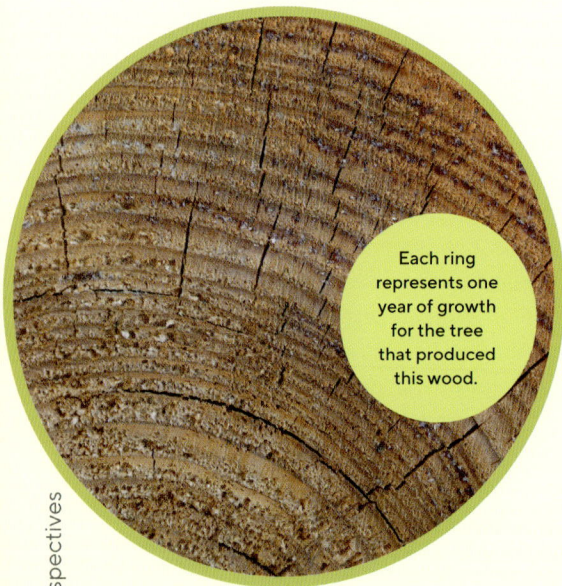

Each ring represents one year of growth for the tree that produced this wood.

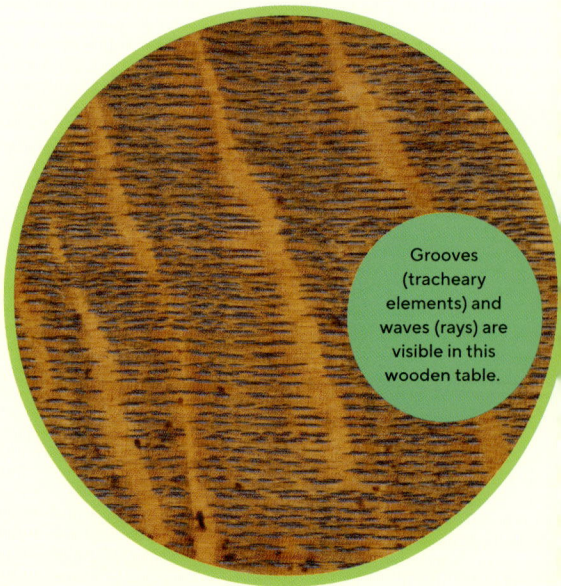

Grooves (tracheary elements) and waves (rays) are visible in this wooden table.

of these tubes across a slice of wood are what give it its characteristic "tree rings." In certain species, large tracheary elements are produced each growing season early in the year, and they narrow as the summer goes on. This pattern, year after year, leads to the rings we count to age a tree or our furniture.

The long waves in wood are cell aggregates called rays and are oriented perpendicular to the stem axis. Their orientation relates to their function—they help shuttle nutrients between the inner and outer portion of the tree. Rays and tracheary elements come in many different shapes, sizes, and orientations that are distinct to each species. Artists, crafts-people, and woodworkers take advantage of this variation to construct the beautiful and functional wooden objects of our everyday lives.

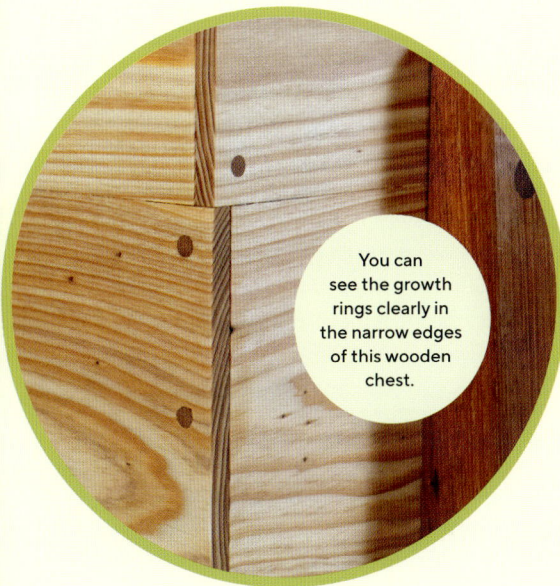

You can see the growth rings clearly in the narrow edges of this wooden chest.

NO SEASONS? NO GROWTH RINGS.

If your coffee table or nightstand is made out of tropical wood, it may not contain growth rings. This is because seasonal rings only occur in seasonal habitats. In the ever-wet tropical forests, growth rings are often absent.

83

Imagine the life of a plant growing in a sidewalk crack, from seed to death.

As Tupac Shakur wrote, "Long live the rose that grew from concrete when no one else ever cared."

Every day, in cities and towns across the world, millions of people step over, around, and right on top of resilient plants that germinated and grew in sidewalk cracks. These can be plants that have grown from seeds that were wind-dispersed, scattered and kicked around, or carried off by ants. However they ended up wedged in a sidewalk crack, the only soil they have—if you can even call it that—is the loose lint that fell out of your pocket, or the dust kicked up from your shoe. Those gritty seeds received the right combination of conditions (the right amount of light and water and the right temperature) to germinate there and attempt a shot at life—isn't that what we are all trying to do?

Many of these plants will have faced the challenge of sending roots down, around, and past the sidewalk toward sources of water and nutrients. Maybe some of them never did send roots past the concrete layer, tolerating boom and bust cycles of water availability and a precar-

ious anchor to their spot on the globe. Water becomes scarce on the sidewalk not because it rains any less there but because soil, which sidewalks so often lack, acts as a reservoir that keeps water accessible long after the rain has stopped, and concrete does not.

The challenges of living as a sidewalk plant come not only in the form of scarcity but also overabundance. Salts and minerals that occur in low concentrations in rural environments are often present in high concentrations in cities—especially in cold regions where salts are sprayed seasonally to melt the snow and ice. Water is drawn to salts, making it more difficult for plants to pull that precious water from salty substrates, and once they enter the plant body, salts disrupt the plant's internal regulation of water. In spite of all this, the plants that grow in sidewalks often flower, fruit, and disperse seeds to start a new generation of hardy survivors.

Common purslane (*Portulaca oleracea*) and cleavers (*Galium*) find a home between a sidewalk and road.

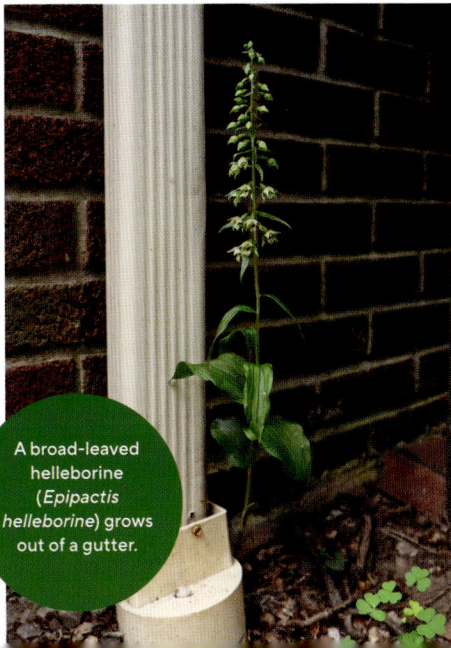

A broad-leaved helleborine (*Epipactis helleborine*) grows out of a gutter.

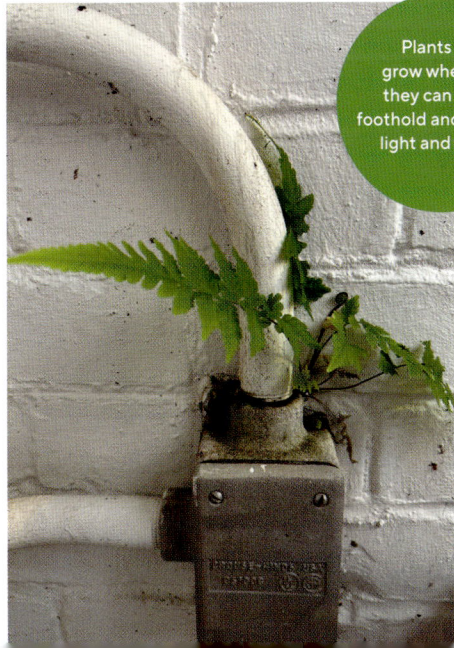

Plants will grow wherever they can find a foothold and enough light and water.

84
Count the number of other species living on and around one tree.

Life follows a latitudinal gradient, with most species occurring closer to the equator, like in the tropical rainforest of Panama.

Plants form the energetic foundation of nearly every ecosystem on our planet.

They are producers that generate sugar via photosynthesis to fuel themselves, whereas animals, fungi, and many microbes are consumers that must eat producers or other consumers to survive. Thus, it is common to find diverse communities living on and around the plants they eat, which in turn attracts more organisms that feed on the plant-eaters.

Insects, and beetles in particular, are by far the most diverse group of animals, and it is not a coincidence that many of them live intimately plant-dependent lives. In addition to their crucial energetic role, plants also represent the overwhelming majority of all biomass on land. Most of the living matter on dry land (over 82 percent) belongs to a plant. Some long-lived trees are the heaviest individual organisms alive today (see prompt 91). The enormous bodies of trees create ample surface area, often with complex nooks and crannies that are habitat for other species, including other plants. And because trees are long-lived, the habitats they create support these other life-forms through many generations.

Resurrection fern (*Pleopeltis michauxiana*) grows on a branch of a southern live oak tree (*Quercus virginiana*), creating habitat for other organisms.

Tree trunks
bend and sway
in the wind.

85
Listen to the plants.
What sounds do you hear them making?

Individual plants are sessile organisms, meaning that once they are rooted in place, they will never move from that spot. However, even though they can't move their whole bodies from place to place, plants are not stiff or static. They sway, shake, and flutter when forced by wind, water, or other organisms, and the sounds these movements produce—rattling branches, groaning trunks, rustling leaves—remind us of their flexibility.

Of course, plants don't intend these movements in the way we intend a high five or a wave, but in an evolutionary fitness sense, the swaying of tree trunks and branches does serve a purpose. Just as skyscrapers are designed to sway in the wind, imparting resilience, trees flex to absorb the force of the wind without snapping.

The sound of rustling leaves is another reminder of the wind's influence on plants. The kitelike leaves need to hold fast while they flap and flutter. In deciduous trees, leaves maintain a solid grip on their branches throughout summer thunderstorms only to pop off and float gracefully to the ground in the fall. The attachment point between a leaf and its branch is called the abscission zone (see prompt 29).

Bamboo stalks clack together like wind chimes.

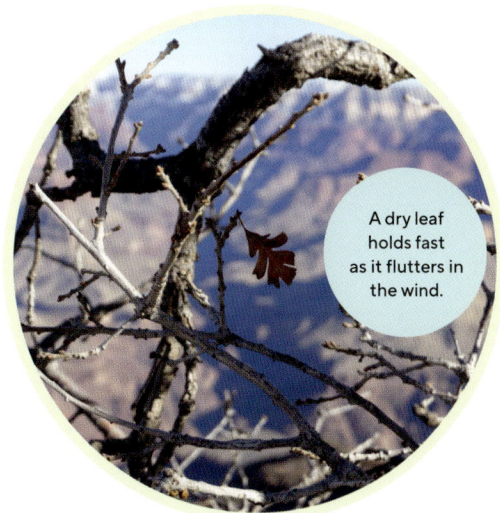

A dry leaf holds fast as it flutters in the wind.

86
Can you find multiple species of plants in the same family?

Draw a small family tree of you and your parents. Now, extend that back to their grandparents, and their grandparents before them. If we continued this exercise far enough back, the ending point or common ancestor would be the first modern human. Keep going, and it would be an apelike creature. Continue even further, and eventually you would get to the first single-celled life on Earth, roughly 4 billion years ago.

Just as within our own families, all living things on Earth are related to each other. This, of course, is true with plants. Scientists often like to put things in groups to make sense of them. We do this with nonliving

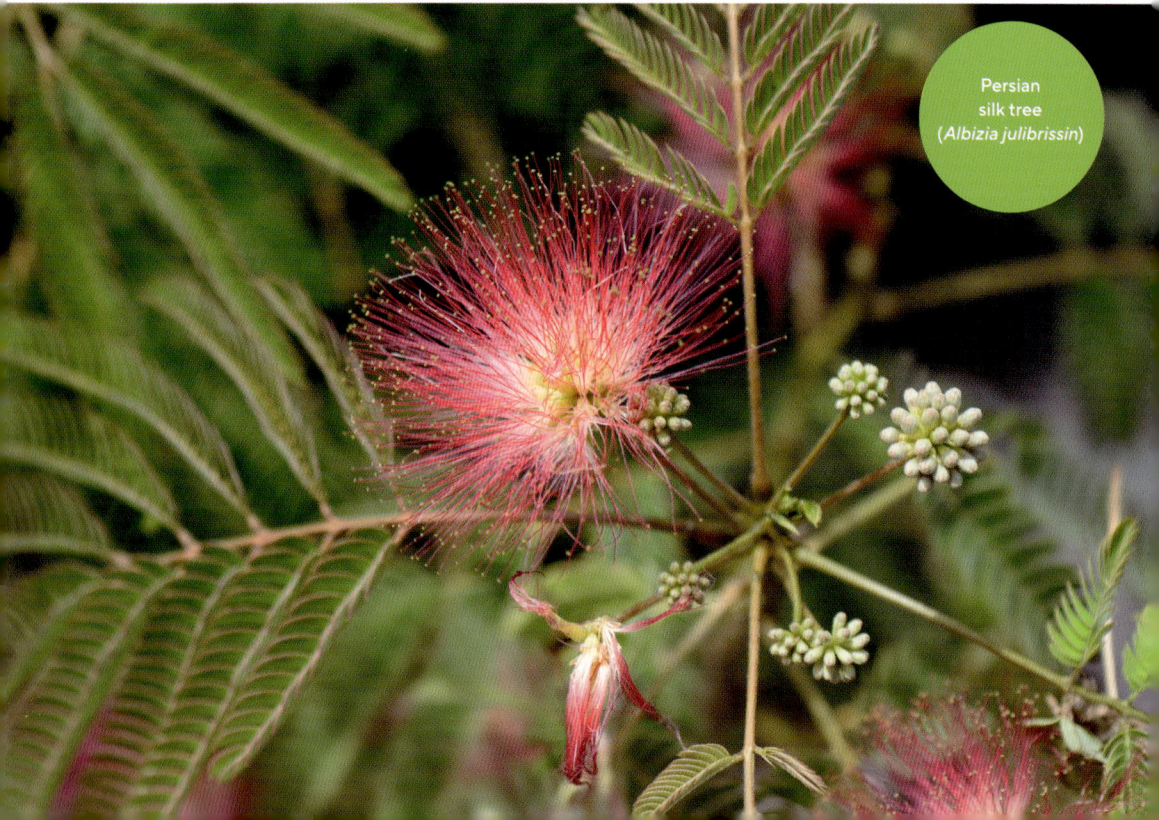

Persian silk tree (*Albizia julibrissin*)

organisms like rocks and gemstones as well as with species in the living world. We group living organisms into kingdoms, phyla, classes, orders, families, genera, and species. (You may have had a specific mnemonic from high school biology. Ours was King Philip Came Over For Good Soup.) These groupings, while not fully reflecting the true complexity of nature, help us understand and make sense of biodiversity and evolutionary connections.

A useful grouping of plants are families, which in plants end in the suffix -aceae, pronounced AY-see-ee. There are hundreds of plant families, including the rose family (Rosaceae), the bean family (Fabaceae), and the mint family (Lamiaceae). Some families, like the mints, have thousands of species, while others, like ginkgo (Ginkgoaceae) have a single species—Ginkgo biloba. Each family is often characterized by the traits they possess. The magnolias (Magnoliaceae) tend to have simple and entire leaves, while the hickories (Juglandaceae) often have dissected leaves. Just like we can tell who our relatives are by our shared traits—you may share the same eye color, hair color, nose, or ears as your siblings—a walk in the woods can reveal which plant species are related to others.

Eastern redbud (*Cercis canadensis*)

Scotch laburnum (*Laburnum alpinum*) flowers

All three of the species pictured are in the bean family, Fabaceae, which means they produce a unique type of fruit called a legume. Legumes develop from a single carpel, are typically dry at maturity, and split open along two seams to release their seeds.

87
Has a plant hurt you recently?

Many plants have evolved to attract some types of animal attention (like pollinating flowers or dispersing fruit) and repel others (like herbivory or trampling). Since plants can't run away, their defenses must work passively. That is, an animal has to initiate contact with the plant to encounter the defense. Mechanical defensive strategies, like spines, thorns, and prickles, are visible at a distance and are presumably easier for animals to learn to avoid. A cactus looks like a plant that should be approached with caution (see prompt 27).

Chemical defenses, on the other hand, may be harder to recognize for animals. Folks living in eastern North America know to avoid poison ivy (*Toxicodendron radicans*), but the plant doesn't boldly advertise its ability to inflict an itchy, painful rash—indeed, that is something learned, perhaps the hard way. Some plants with chemical defenses only need to be gently brushed to deliver an uncomfortable experience, while others, like cow parsnips (*Heracleum maximum*), need to be physically damaged to release their toxins.

TOP: Poison ivy (*Toxicodendron radicans*) is famous for the rash-causing oils on its shiny leaves.

MIDDLE: A Panama rubber tree (*Castilla elastica*) oozes latex, which can gum up an herbivore's mouth when it takes a bite.

BOTTOM: Hawthorn (*Crataegus*) has large, sharp thorns to deter large herbivores.

When the leaves or stems of cow parsnip are broken, a clear sap is exuded that has phototoxic properties. If the sap comes in contact with skin, it causes painful blisters upon exposure to sunlight (specifically UV wavelengths).

ATTRACTIVE REPELLANT

Interestingly, some chemical defenses—like caffeine and menthol—actually attract the attention of humans even though they likely evolved to repel insect herbivores. We use these in our daily lives and probably don't give much thought to why the plant evolved to produce them (see prompt 75).

88

Notice the feeling when you touch a plant.

Think of a time when you waded through a field of grass reaching to your knees, or pushed through a thicket of shrubs in the forest understory. How about a time when you ran your hand through the stubble of a freshly mowed lawn, or when you brushed aside a wiry tree branch at face height during a hike? When we come into contact with plants, how often do we stop to recognize that we are touching other living organisms?

Most of the plant tissue we touch, from leaves to fruit to flowers, is alive. Although plant cells are stuck in their relative positions due to their strong cell walls, many plant structures are remarkably flexible. Plants can flex but bounce back to a form defined by the matrix of their cell walls. This means that when we push on plants, they push back, reminding us that they are alive, too. They assert their poses, reaching toward light and water, away from strong wind, resilient to the brush and bump of our animal limbs.

Feathery inflorescences of wavy hair grass (*Deschampsia flexuosa*) are soft to the touch and bend in the breeze.

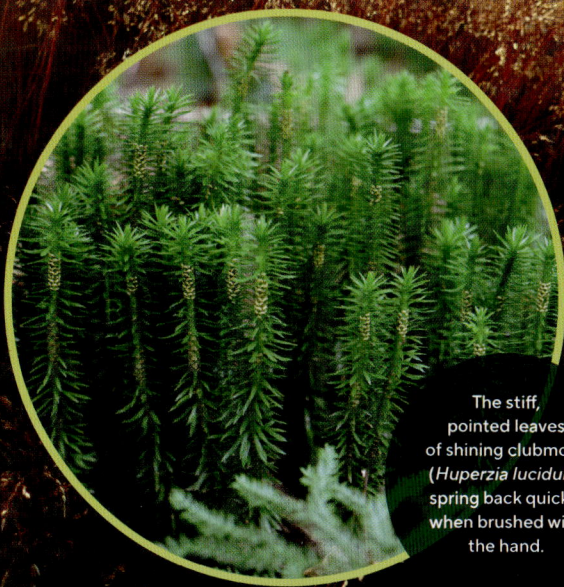

The stiff, pointed leaves of shining clubmoss (*Huperzia lucidula*) spring back quickly when brushed with the hand.

Food plants are often grown near the equator so that seasonality in their wild habitat no longer determines when we have access to fresh fruits like apples, strawberries, or blueberries.

89
Think about the plants you care for and those that care for you.

We directly care for our houseplants, gardens, agricultural fields, parks, and more. In these contexts, we might provide plants with water and access to sunlight, offer fertilizer rich in nitrogen and other nutrients, and take steps to remove or guard against herbivores and pathogens. There are thousands of years of caretaking history between humans and plants from cultures around the world. The plants we care for are ones that we touch, ones that we directly lay eyes on. However, many of the plants that provide for us live their lives far away from ours.

Most global of all are the plants that continually replenish the oxygen we breathe. As plants and cyanobacteria around the world perform photosynthesis, they release oxygen molecules into the air. These molecules travel about 1,000 miles per hour, meaning they could circle the earth in approximately one day. Their movement is much more random than a straight line around the globe, but it is no stretch to imagine that the oxygen released by a single plant just one year old may have already spread around the entire planet.

Wood for building homes and furniture is shipped great distances before being used.

90
Describe the movements of a tree.

Trees may seem to stand perfectly still, but in fact their bodies are constantly in motion. Some plant movements even occur on animal-like scales of time and space. For example, we can easily track the steady gyrations of tree branches in the breeze or the spiraling descent of winged maple fruits with our naked eye. Since plants flex and bend under outside forces, some of their movements, like the frantic flapping of leaves in a strong gust of wind, are much faster than our muscle-powered animal bodies are capable of.

Of course, plants also move at scales of time and space that we can't easily observe with the unaided eye. On a microscopic scale, the stomata on a tree's leaves open and close every day in response to environmental factors like sunlight and water.

Other movements occur at the macroscopic scale but at rates of speed too slow to easily observe. A new leaf emerging in spring may grow thousands of times larger in volume, but this process plays out over the course of days to weeks and is best observed via time-lapse photography. Plants also grow or turn toward the sun, a response called phototropism.

Stomata

Each individual stomate is a tiny valve enclosed by two cells, and the hundreds or thousands of these minuscule moving structures on each leaf enable a plant to continuously update its physiology to its current environment.

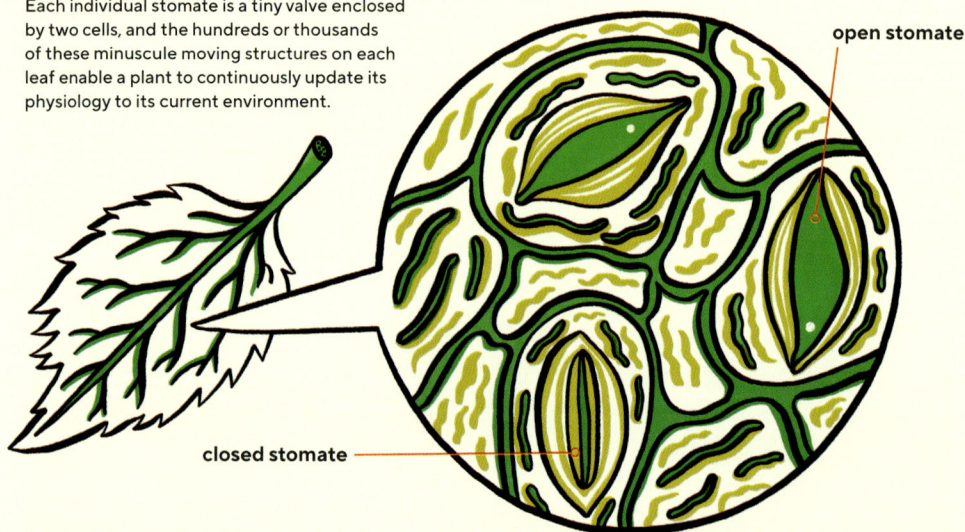

open stomate

closed stomate

Plants are always in motion, but sometimes we need to think at different scales of time and space to perceive all their movements.

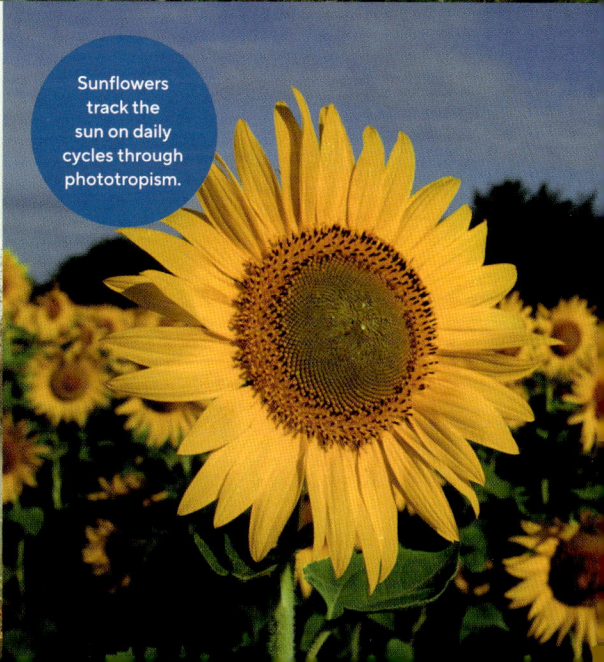

Sunflowers track the sun on daily cycles through phototropism.

A staggeringly tall red oak reaches to the sky.

91
Try to guess how much a tree weighs.

The first law of thermodynamics states that energy is neither created nor destroyed. The sun is a nuclear reactor that slams atoms together and, in the process, releases energy that beams down onto Earth. This energy is unusable by most life on our planet. However, plants have figured out how to harness it. They use the energy of sunlight to run a microscopic power plant where the output is sugar. This impressive evolutionary invention is, of course, photosynthesis.

During this process, plants use the power of photons from the sun to break

water molecules and use the released energy to power cellular machines that turn carbon dioxide into sugar. The carbon dioxide that plants use in this process is from the air. It is not an intuitive concept, but they build their bodies by breathing. All of the mass in the tallest tulip poplars and the oldest pines are literally pulled from thin air. They are the original body-builders, but instead of bulking up on pasta and rice, a tree can easily amass 20 tons from pure gaseous carbon dioxide.

The energy in all that rice and pasta comes from somewhere, though. Since energy cannot be created, all the energy we consume comes through plants, from the sun. Whether we eat plants directly or eat animals that eat plants, it just about all comes back to plants.

The massive trunk and branches of this oak tree were built entirely by breathing air and utilizing the carbon in carbon dioxide to build carbohydrates.

Tropical rainforest ecosystems, like this Panamanian forest, can harbor roughly twice the plant biomass of temperate forests.

perspectives

92

Can you find an herbaceous plant that is taller than you?

Tall blue lettuce (*Lactuca biennis*) can grow up to 12 feet tall.

While most flowering plants are capable of producing secondary growth (or wood) from a layer of cambial tissue in their stems, many thousands of species never grow significant woody stems. These non-woody plants are called herbaceous—basically, plants that are not trees or shrubs.

Wood is an incredibly strong and surprisingly flexible material, which allows some trees to reach heights of 200 feet or more, making them by far the tallest organisms ever to live on our planet. While herbaceous plants never approach these great heights, they can still grow tall enough that holding their bodies up against the force of gravity is not a trivial engineering challenge.

Herbaceous plants, which typically only live for one year or die back to their roots each year (see prompt 93), often solve this problem by producing hollow stems or stems filled with **pith**. These stems utilize the cylinder, a very strong shape, to maximize sturdiness for a minimal input of body mass. With only one growing season to rise from ground level, a species like smooth cordgrass (*Sporobolus alterniflorus*) can reach nearly 5 feet tall, and some species like Japanese knotweed (*Reynoutria japonica*) can reach 10 feet tall by growing large, lightweight, hollow stems.

Tall goldenrod
(*Solidago canadensis*)
grows to about
three feet high.

Japanese
knotweed supports
its height with large,
hollow stems.

93
Find an herbaceous plant. Did it exist last winter?

The range of plant lifespans is immense. Many live for less than one year, while others, like the Great Basin bristlecone pine (*Pinus longaeva*), can survive for millennia (see prompt 95). It is fairly obvious that large plants like trees live for many years. However, guessing the lifespan of smaller, herbaceous plants is not always so straightforward. Herbaceous plants may live for one year (**annuals**), two years (**biennials**), or longer (**perennials**); it is not always possible to tell from a single observation which of these categories applies to a given plant.

The size of the leaves, the toughness or softness of the stems, and the height of the plant are all factors that can't be trusted for inferring lifespan. Sometimes, though, the underground portions of a plant can provide a clue. The aboveground portions of most herbaceous perennial plants die back significantly for part of the year no matter how long-lived the species. This typically occurs during winter in the temperate region (if temperatures get cold enough) or during the dry season in the seasonally dry tropics. Often, no trace of the plant will be visible aboveground during this time, but in perennial plants, a root system or **rhizome** (which is an underground stem, not a root) may survive.

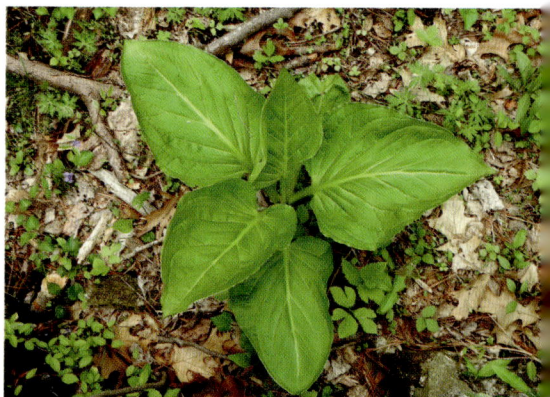

Skunk cabbage (*Symplocarpus foetidus*) is a perennial with a long-lived rhizome.

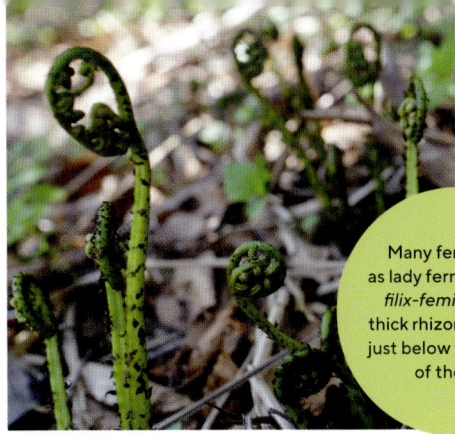

Many ferns, such as lady fern (*Athyrium filix-femina*), have thick rhizomes sitting just below the surface of the soil.

Biennial plants often form a tight cluster of leaves that lie close to the ground (called a **basal rosette**) in their first year before sending up a taller stem that produces flowers in year two. A good example is the common evening primrose (*Oenothera biennis*). Biennials are best identified by looking for that basal rosette and returning to observe the same individual plant next year.

Of course, for all plants, including annuals, some precursor of the plant did exist in the previous year, whether it was a seed or spore lying dormant in the soil or a cell in the flower of its mother plant, awaiting pollination.

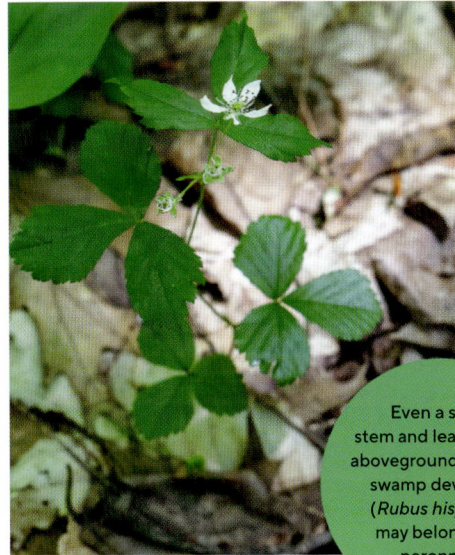

Even a small stem and leaf system aboveground, like this swamp dewberry (*Rubus hispidus*), may belong to a perennial.

The common evening primrose (*Oenothera biennis*) has a Latin species name referring to its biennial lifestyle.

Some plant communities are not densely packed, like this community of succulents in a desert. This is often the case in especially harsh environments where resources like water are scarce.

94
Sit down somewhere outside. How many individual plants are within arm's reach?

Communities in temperate, wet forests where resources are abundant are often densely packed, like this patch of ephemeral wildflowers.

Plants often grow in dense communities,

usually in direct contact with their neighbors, which leads to competition. Limited resources, like access to water and sunlight, affect the number and type of plants living in a certain area. As shrubs and trees grow tall, they intercept sunlight before herbaceous plants beneath them and can shade them out of a thriving existence. Where grasses spread thick layers of fibrous roots, it can be hard for other seedlings to get enough water to establish themselves and grow.

But another way of framing all of this competition is as coexistence. Plants are able to live together in remarkably dense and diverse communities. In part, this is because the main resources they require—water, sunlight, and air—are often abundant. The plant communities we observe today assembled through ecological and evolutionary processes. They are composed of species with variations in traits like root depth and shade tolerance, which allow them to coexist. Over evolutionary time, they have filled their own niches within the space they have.

95
Consider an individual tree.

What did this landscape look like when that tree germinated from a seed?

An old bristlecone pine (*Pinus longaeva*) grows in the White Mountains of California.

When we think of long-lived organisms, we often think of turtles, whales, and the Greenland shark. However, some of the oldest living organisms on Earth are trees. In fact, to the extent of human knowledge, the oldest non-clonal living organism on planet Earth is an unassuming pine tree—the bristlecone pine (*Pinus longaeva*), to be exact (for clonal species, see prompt 72). At an estimated 5,000 years old, this individual tree germinated when camels were first domesticated in Egypt. It was 2,000 years old when Jesus was born, and 3,000 years old when the Vikings navigated the North Sea to invade England. At 500 years old (nearly 4,500 years ago), it was already older than the oldest living vertebrate ever identified (a 480-year-old Greenland shark).

These are the longevity champions, but every tree has experienced some window of history as the landscape changed around it. A patch of young forest with a single large tulip poplar (*Liriodendron tulipifera*) in the middle may signal that the forest used to be a field. A dense forest of massive trees, like the Darien Gap between Central and South America, suggests the forest has not been cut in centuries. Likewise, large sugar maples forming a straight line through a northeastern US forest can signal the edge of a previous farm or roadway.

Just like maps tell us ways to navigate through space and roads give us the paths to take, trees can teach us the history of a landscape.

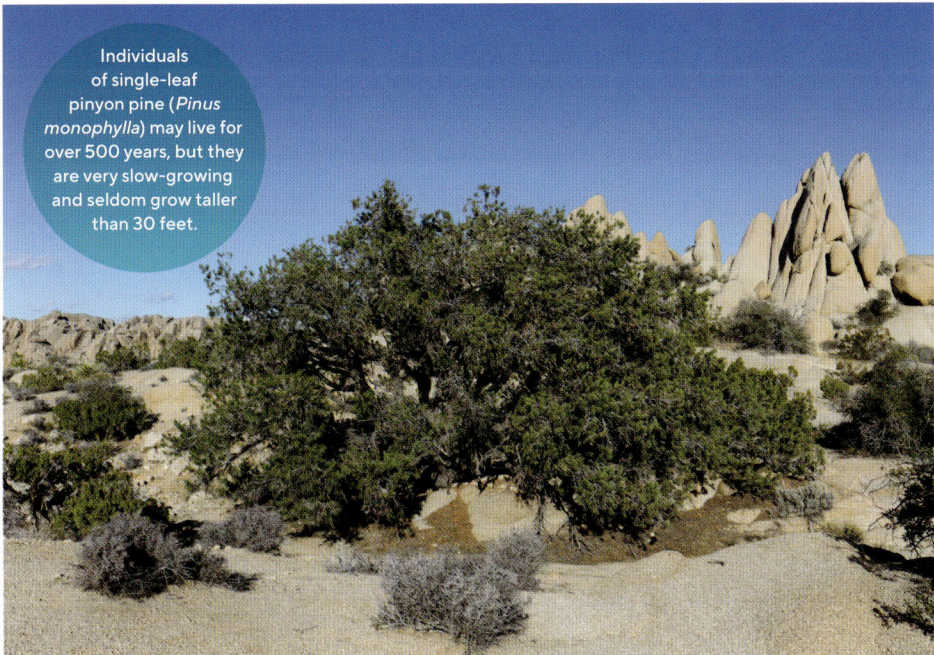

Individuals of single-leaf pinyon pine (*Pinus monophylla*) may live for over 500 years, but they are very slow-growing and seldom grow taller than 30 feet.

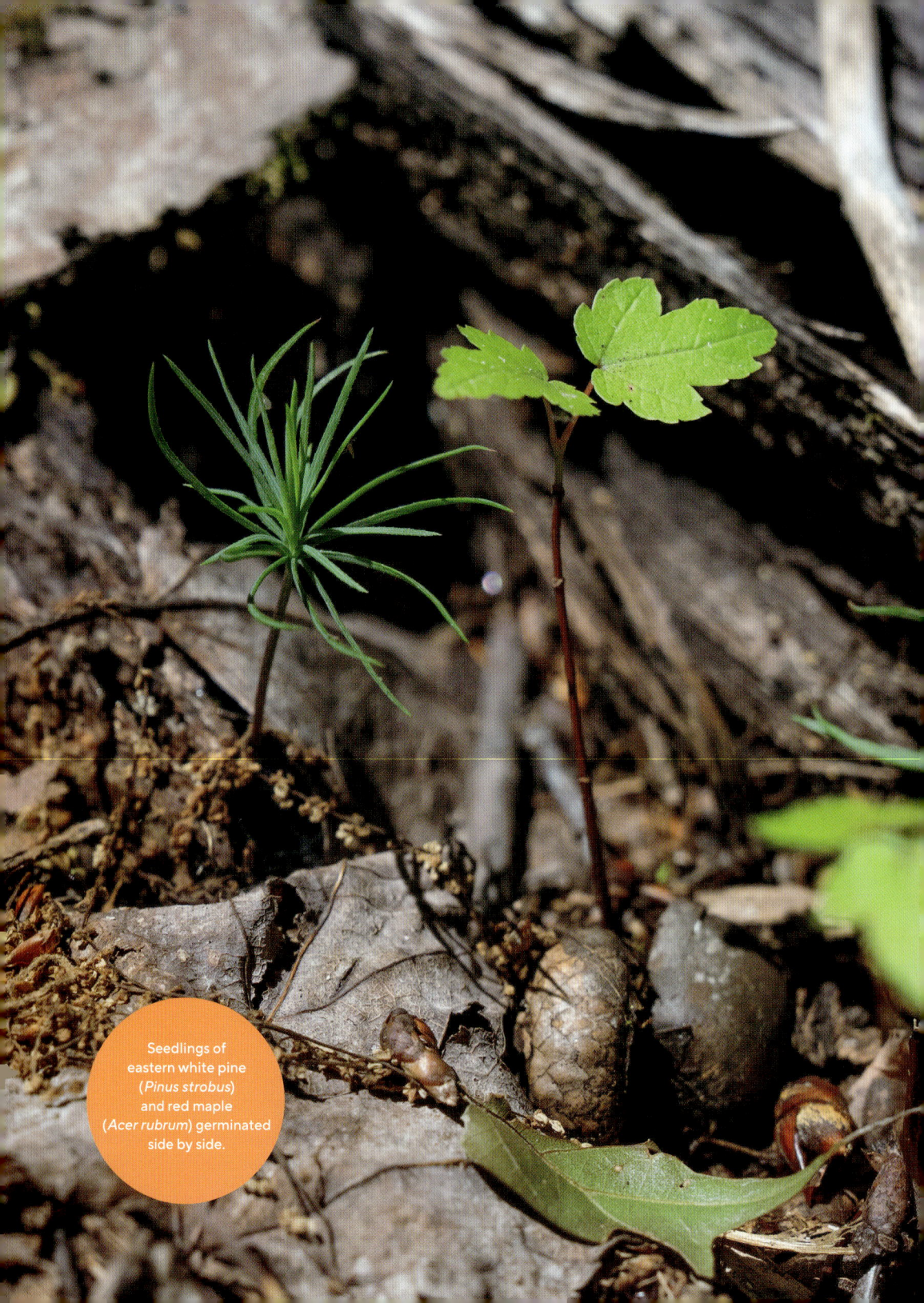

Seedlings of eastern white pine (*Pinus strobus*) and red maple (*Acer rubrum*) germinated side by side.

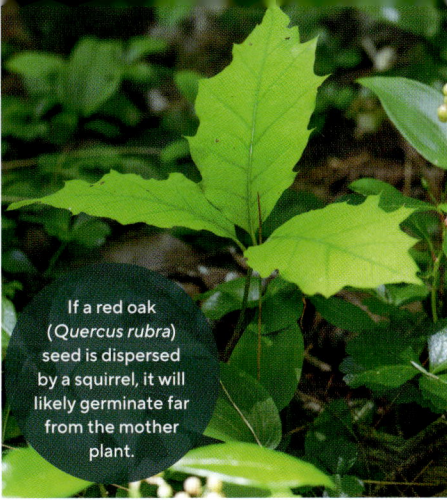

If a red oak (*Quercus rubra*) seed is dispersed by a squirrel, it will likely germinate far from the mother plant.

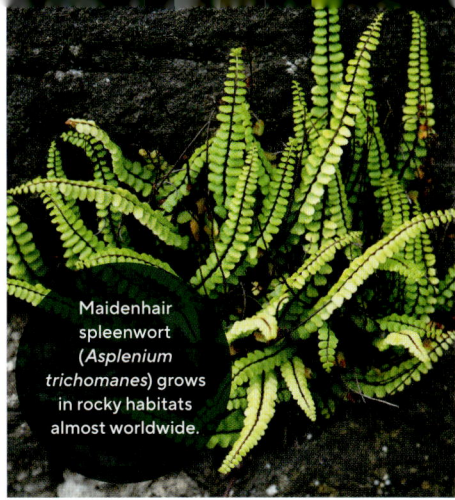

Maidenhair spleenwort (*Asplenium trichomanes*) grows in rocky habitats almost worldwide.

96
Find a plant and try to guess which individual may have been its mother.

As seeds, plants can be very mobile. Knowing what the fruit of a flowering plant (which contains the seeds) looks like can give a strong hint as to whether its seeds tend to disperse far away from their mother plants or settle right next door. For instance, those that are fleshy are often animal dispersed, while those that are indehiscent and winged may disperse with the wind (see prompt 53). Fruits that can catch a ride on the wind or be eaten by a bird tend to be carried far distances. Even elephants have been shown to move seeds many kilometers.

Outside of the flowering plants, groups that disperse by spores, like ferns and mosses, can travel enormous distances if they get swept up into the stratosphere (see prompt 69). We know this can happen because some fern species like black spleenwort (*Asplenium adiantum-nigrum*) have recently expanded their range to multiple continents. By spreading out through seed and spore dispersal, new plants can avoid directly competing for resources like water and light with their already-mature parents. Leaving more space between individual plants can also help alleviate herbivore or pathogen pressure by reducing the likelihood of outbreaks.

perspectives

Cultivated rose

Wild rose

Double-flower wild rose

97
Compare a wild rose flower to a rose in a florist's bouquet.

How many petals does a rose flower have? That's a bit of a trick question, because it depends on whether the rose in hand is from your local flower shop or a rose in the wild. Wild rose flowers have exactly five petals (see prompt 74). A rose encountered in a bouquet probably has 15 or more petals crowded so closely together that they might be difficult to distinguish. Wild peonies have 10 or fewer petals, but the peonies in our gardens seem to have too many petals to count. The same can be said of wild versus cultivated buttercups (*Ranunculus*) and many other commercially grown flowers.

Over and over again, we have found mutant plants (sometimes called sports in horticultural lingo) that bear flowers teeming with far more petals than average, and we have chosen to keep and propagate

In this double-flower rose mutant, the stamen whorls have turned into petals.

them for their unusual appearance. These are often referred to as "double flower" varieties and have been written about since Theophrastus, a pupil of Aristotle, in the third century BCE. Biologically, double-flower mutations transform other floral organs into petals (most often stamens, but sometimes carpels as well). A surprisingly simple genetic mutation can bring about this seemingly drastic transformation, and because the genetic mutation is so simple, it has now been found in hundreds of different flowering plants.

Ironically, many of these double-flower mutants are now the only versions of plants that most people are aware of. For instance, roses, peonies, and carnations can rarely be purchased with their natural petal numbers. Originally prized for their uniqueness compared with more common wild forms, these mutants are now the basis for our cultural archetypes. Both forms (wild and double flowered) are beautiful in their own ways, but they are more fascinating when they are considered together.

A mutant flower of an orange daylily (*Hemerocallis fulva*) shows stamens that have mutated into petals.

Leaf litter plays a crucial role in nutrient cycles, returning carbon and nutrients like nitrogen and phosphorus to the soil to be used again by other plants, fungi, and microbes.

98
What do you exchange with plants?

With few exceptions, Earth is a closed system with a finite pool of atoms. These atoms come together and move apart, combining and recombining into myriad chemical compounds, but they are not created or lost. Rather, these elements cycle through our planet's ecosystems. These cycles of carbon, nitrogen, oxygen, and other elements represent the major recycling system of life. Some of these cycles are straightforward, like the oxygen cycle: Oxygen is breathed in by living things from

Grassland ecosystems are responsible for about 20 percent of all carbon fixation on land, releasing large amounts of oxygen in the process.

the atmosphere, then breathed out as carbon dioxide, which is absorbed by plants and released back into the atmosphere as oxygen.

Other cycles, like nitrogen, may be more complex, involving lightning, power plants, and bacteria. Plants play a central role in each of these chemical cycles involving the building blocks of life. This is in part because of their abundance and in part because of the alchemy of photosynthesis. The oxygen atoms we inhale with every breath are sure to have passed through plants (or oceanic plankton), and the carbon dioxide we exhale will feed back into their photosynthesis reactions. The atoms that make up our bodies are only here temporarily. Perhaps they have just come from visiting a plant, or perhaps that is where they will end up next. Indeed, when we decompose, our carbon and nitrogen molecules will likely be transferred to some bacteria and maybe eventually into a plant, if we are lucky.

perspectives

99

Place a leaf in a plastic bag for a day.

When you come back, does the bag have condensation? Where does the water taken up by a plant's roots go?

What happens when you breathe into a plastic bag?

The bag inflates, of course, but you'll see something else happen: It gets humid or foggy. This is because the gaseous waste we produce when we exhale includes water vapor that fills the bag with molecules that condense on the surface of the plastic.

Plants also breathe. They are living, after all. They have mitochondria, which require oxygen to power their bodies. While they do expel carbon dioxide and water vapor as byproducts of respiration (see prompt 100), the majority of water expelled from leaves is through a different mechanism. For plants to take up carbon dioxide, they must keep the tiny pores (stomata) in their leaves open. When this happens, water freely evaporates from the air spaces inside a leaf. Given the amount of carbon dioxide in the atmosphere (about 400 parts per million, relatively low compared to earlier in Earth's history,

and about 10 times lower than the average concentration of water vapor in the air), immense amounts of water are evaporated out of these stomata for minuscule amounts of carbon dioxide. In fact, nearly 400 water molecules are lost for each carbon dioxide molecule gained.

This water loss from the leaf actually drives water movement through the plant, and it's called **transpiration**. Think of each water molecule as a link in a chain. As each molecule evaporates out of the stomate, it pulls another molecule up through the plant. This link of molecules stretches from the tip of the canopy all the way down to the soil. So when we place a leaf inside a plastic bag, whatever water molecules are left in the plant will evaporate out and eventually condense on the inner surface of the bag, leaving behind dew droplets— the physical reminders that plants breathe, too.

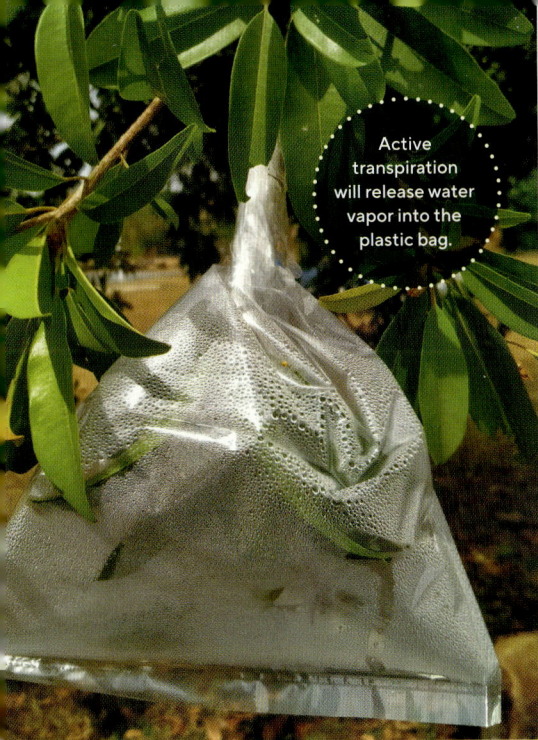

Active transpiration will release water vapor into the plastic bag.

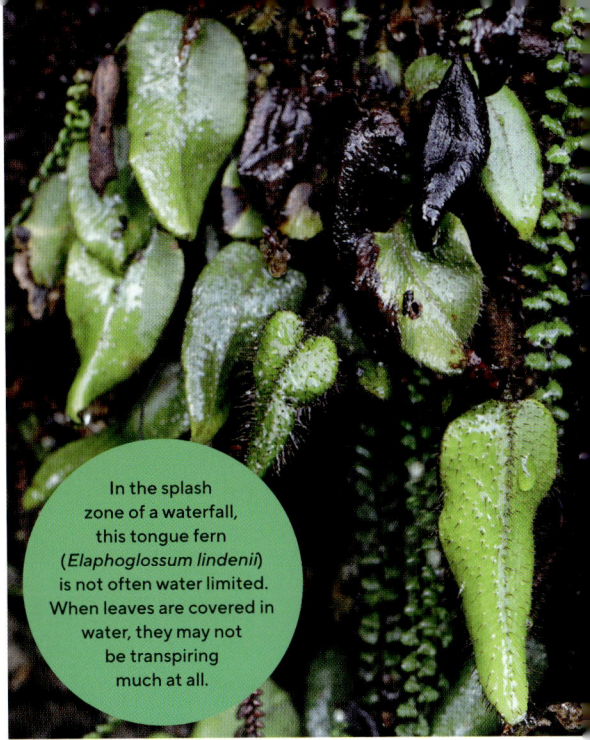

In the splash zone of a waterfall, this tongue fern (*Elaphoglossum lindenii*) is not often water limited. When leaves are covered in water, they may not be transpiring much at all.

Transpiration

Water loss through leaves pulls more water up through the roots.

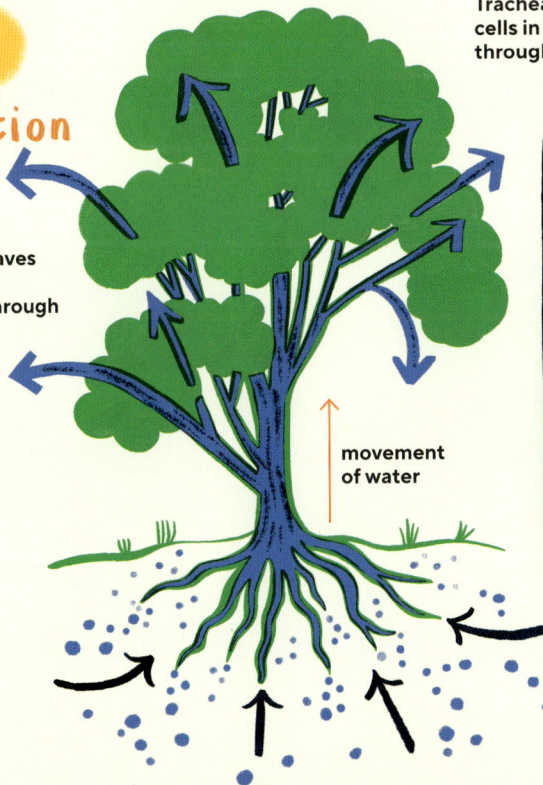

movement of water

Tracheary elements are tubular cells in the xylem that move water through the plant's body.

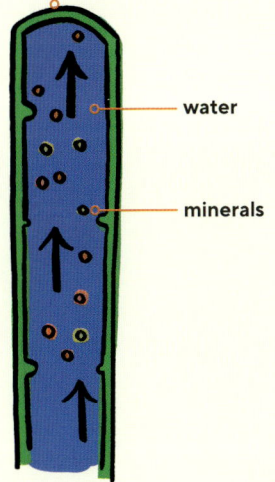

water

minerals

Some trees that live in very wet habitats, like mangroves (*Rhizophora*), produce roots that reach partially above the water, which is thought to aid in gas exchange in addition to stabilizing the plant in wet and shifting substrates.

100

Can you see the structures plants use to breathe?

HINT: You may need a hand lens to find them!

Animals, like us, breathe in oxygen and expel carbon dioxide, and we think of plants as doing the opposite. However, plants also need to breathe oxygen like animals. They do so to perform the same chemistry of **cellular respiration**—turning sugar molecules into usable energy in the body. In the process, they produce carbon dioxide as a byproduct. In the leaves, stomata are a two-way road for gases (see prompt 90). Stomata, though miniscule, are crucial sites of gas exchange and are typically concentrated on the undersides of leaves.

However, these are not the only structures plants have evolved to breathe. Other parts of the plant sometimes develop specialized features to aid in absorbing and expelling oxygen or carbon dioxide.

Many different species of trees produce pores in the bark called lenticels that aid in gas exchange. Root tissues also perform cellular respiration and require a consistent influx of oxygen. In well-draining soils, roots have access not only to water but also to small pockets of air in the spaces between soil particles.

The bark of a paper birch (*Betula papyrifera*) shows its lenticels. Birches and cherries (*Prunus*) often bear large, horizontal lenticels, especially on young stems and branches.

A Chinese plum (*Prunus salicina*) has peeling bark with large, lateral lenticels.

101
Express gratitude toward an individual plant.

Relating to individual plants starts with careful noticing.

Direct your attention to one particular plant. Maybe it's a plant with special significance to you, like a houseplant, a favorite tree on your street, or something you planted in the garden. Maybe it's a plant you're just encountering for the first time. Take a moment to observe it closely. What do you appreciate about this individual plant?

Focusing our attention on an individual human is something we do often. Turning that same sort of attention toward a plant is likely less familiar. Now comes the greater challenge: How can you communicate gratitude from one species to another?

Using language is technically an option, but it will be much more meaningful to you than to the plant. Across species boundaries, communicating through action is more effective. Can you perceive any stresses or threats to this plant? How might you be able to improve its health and help it thrive?

Return to your observations. Do you notice any visual signals like wilting leaves thirsty for water, or elongated stems reaching out for more light? Through a little knowledge of plant needs and some careful observation, we can receive messages like these from plants. How can you communicate back?

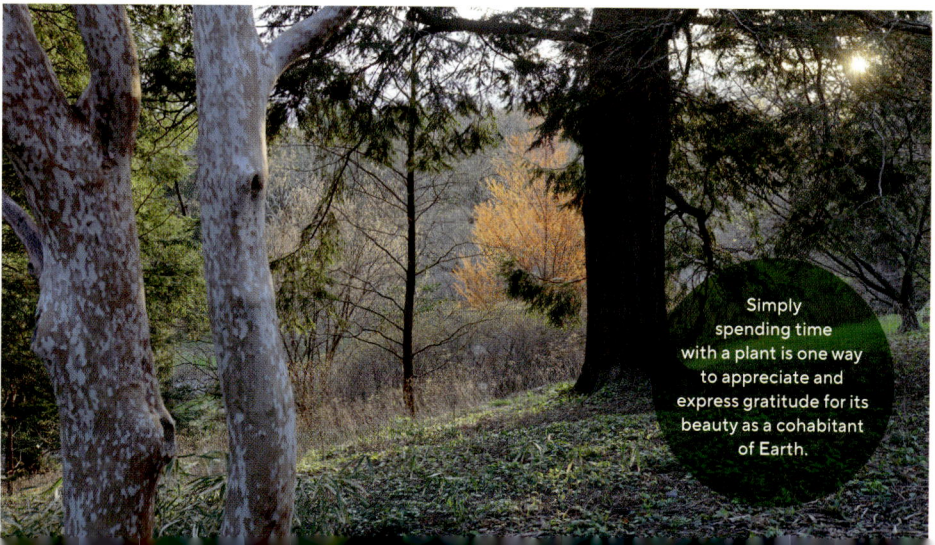

Simply spending time with a plant is one way to appreciate and express gratitude for its beauty as a cohabitant of Earth.

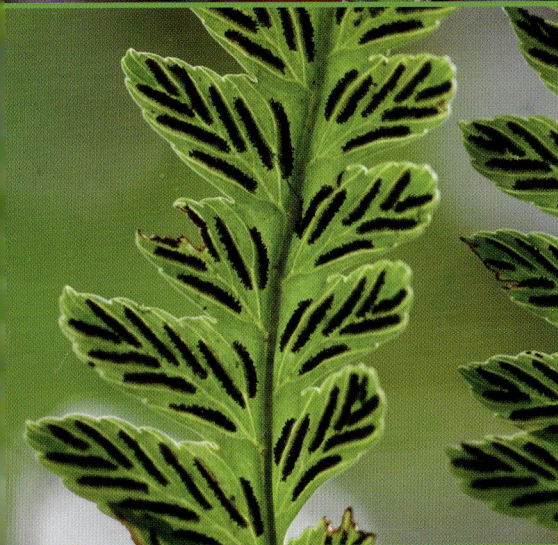

WHAT'S YOUR SPARK PLANT?

A spark plant can be one that is familiar to you. Perhaps it is a flower in your grandfather's garden, a spice your mother cooked with, or a tree you used to climb as a child. We have connections with plants all around us, and we often need only to shift our perspective to reconnect with them. Finding an individual plant that is meaningful to you can provide the spark for you or someone you love to embark on the hobby of botanizing!

Glossary

Abscission zone. The layer of cells at the junction between two structures destined to separate. Often this is between a leaf and its branch. In deciduous trees, the abscission zone remains strong and holds fast during the growing season only to break in the fall at particular points, allowing the leaf to drop off and leave a clean leaf scar.

Achene. A single-seeded dry fruit. For example, sunflower "seeds" including the outer husk are achenes.

Actinomorphic. A radially symmetrical flower (a flower with many planes of symmetry). Contrast with zygomorphic.

Acuminate. Tapering to a long, pointed tip.

Adnate. The fusion of different organs (for example, petals and stamens).

Adventitious roots. Roots produced along the stem or shoot system (as opposed to the root system) of a plant.

Aggregate fruit. A fruit consisting of many loosely fused carpels derived from one single flower (for example, raspberry).

Angiosperm. The most diverse group of plants on Earth. The lineage contains more than 250,000 species. They are defined by the presence of fruits and flowers and make up the overwhelming majority of crops and agricultural goods. Fossil and genetic evidence suggests that they date back to the Cretaceous Period, nearly 150 million years.

Annual. A plant that lives for no more than one year.

Anther. The portion of the stamen (typically at the tip) where pollen is produced.

Anthocyanins. A class of pigment molecules that are produced by plants and responsible for colors from red to purple to indigo. These pigments often underlie red and purple flower petals and the fiery red colors of fall. In addition to producing bright colors, they may protect against excess sunlight as well as herbivory.

Aril. A usually fleshy or nutrient-rich, brightly colored structure that surrounds the seeds of certain plant species. It can be an extension of the seed coat or deposited by the inner walls of the fruit. Arils tend to function in seed dispersal by enticing animals (usually birds) to eat the seeds.

Autotroph. An organism that makes its own food (for example, through photosynthesis like in some bacteria, diatoms, and plants, including algae).

Basal rosette. A tight cluster of leaves produced close to the ground. These are characteristic of plants that produce a burst of growth that then persists through winter, especially biennial plants.

Betalains. A unique class of pigment molecules that are exclusively produced by plants in the order Caryophyllales and produce pink, purple, and red colors. The deep magenta color of a beet is due to betalains.

Biennial. A plant that lives for exactly two years. Typically, biennial plants produce only vegetative growth in their first year before growing reproductive structures in their second year.

Blossom end. The part of the fruit that faces the rest of the flower; where the petals and other floral parts emerge.

Bract. A modified leaf, often located at the base of a flower or inflorescence. Bracts may help to protect the flowers, or they may be colorful and serve to attract pollinators, or they may not have a clear function.

Bryophytes. The lineage of nonvascular land plants that includes the mosses, liverworts, and hornworts. They are the closest living relatives to the vascular plants.

Bud. A structure housing leaf or flower primordia.

Caducous. Describing structures that are ephemeral and fall off after a brief period of time.

Calyx. The outermost whorl of a flower that usually consists of green leafy structures. In many lineages, the calyx is modified for dispersal or protection.

Cambium. The region of stem cells (cells that can differentiate into any other cell type) that gives rise to wood. It forms a cylinder around the stem.

Capitulum. An inflorescence composed of a dense cluster of small flowers, which together appear to form one large flower, that is characteristic of the sunflower family (Asteraceae).

Carotenoids. A class of pigment molecules that produces yellow, orange, and red colors, such as in pumpkins, carrots, and marigolds.

Carpel. The part of the flower that encloses the developing ovules and eventual embryos (after fertilization). It can have an elongated style and stigma, which help receive pollen and facilitate fertilization. The carpel can be solitary (like a peach) or fused (like an orange).

Cellular respiration. The chemical process by which oxygen and sugars are used to produce the molecule ATP, which serves as the energy source for other activities within the cell. Both animals and plants must breathe oxygen and use sugars to fuel this process, which produces carbon dioxide as a byproduct.

Cell wall. The rigid structure of a plant cell that provides support. The cell wall is made out of cellulose, which is a polymer of glucose. This is the defining feature of all plants and is not found in animals.

Charophyte. The group of freshwater algae, including the closest living relatives to land plants.

Chlorophyll. The pigment molecule involved in the light-harvesting steps of photosynthesis. Chlorophyll absorbs red and blue wavelengths of light and reflects green wavelengths, giving chlorophyll-rich plant tissues their green color.

Chloroplasts. Small organelles housed in plant cells. They contain the light-harvesting pigment chlorophyll. Chloroplasts are generally found in the cells of the palisade layer of leaves.

Clade. A group of organisms that includes their common ancestor and all their descendants. For example, the flowering plants (angiosperms) form a clade within the larger clade of plants (Viridiplantae).

Clonal spreading. A growth strategy by which an individual organism produces more individuals without sexual reproduction. These additional individuals may disperse and live separately from the parent, or they may remain connected to the parent. If they remain connected, they are not considered separate individuals but rather a spreading of the single original organism. In all cases, clonal spreading produces individuals (or apparent individuals) that are genetically identical to one another.

Connate. The fusion of multiple organs of the same type (for example, petals fused into a tube).

Convergent evolution. When two or more distantly related lineages evolve a similar structure or function independently. That is, they do not share a common ancestor that possessed the structure or function.

Corolla. The collective word for the petals of a flower.

Costa. The midrib of a fern pinna.

Costule. The midrib of a fern pinnule.

Cotyledon. The first leaf developed from a seedling.

Crenate. A margin with rounded, bumpy teeth (most often describing a leaf margin).

Cuticle. The waxy outer surface of a plant, usually in reference to leaves. The cuticle is predominantly made of cutin, a polymer that is water resistant. This protects the leaf from losing water.

Cutin. A waxy polymer, secreted by the epidermal cells, that makes up the cuticle. It is a major barrier to water and carbon dioxide.

Deciduous. The strategy in some perennial plants of dropping their leaves on an annual cycle.

Dehisce. The process of splitting during the course of development, usually along predetermined seams.

Dentate. A margin with pointed teeth (most often describing a leaf margin).

Dichotomous. Splitting into two branches of equal size.

Dicot. The outdated term given to the flowering plants that produce two first seed-leaves. This refers to all non-monocot flowering plants including the Magnoliids, eudicots, and several other lineages.

Dioecious. An individual plant that bears only pollen-producing or ovule-producing flowers or cones but not both. This is in contrast with monoecious plants.

Diploid. The state of having two copies of each chromosome, one from each parent.

Disc floret. The innermost flowers of a sunflower (Asteraceae) head, usually not producing any showy petals.

Distichous. Leaves held on a single plane in two ranks, one on either side of the stem.

Embryo. A multicellular product of fertilization that can grow into a new individual organism.

Embryophytes. These are the land plants, which include the bryophytes (moss, liverworts, and hornworts), lycophytes (lycopods, spike mosses, and quillworts), ferns, seed plants, and flowering plants. They are called embryophytes because, unlike their algal ancestors, the mother plants retain and nourish their zygotes as they develop into embryos.

Endocarp. The innermost layer of fruit (carpel) tissue, such as the stony pit surrounding the seed of a peach.

Epidermis. The outermost tissue layer of plants. This is the "skin" of the plant.

Epipetric. The life habit of growing on rocks.

Epiphyte. A plant that grows on another plant (but is not a parasite).

Eudicots. The formal name for the clade of plants that is a sister group to the monocots. These species all produce two seed-leaves, or cotyledons.

Evergreen. Maintaining green and functional leaves throughout the year (in contrast to deciduous).

Exocarp. The outermost layer of fruit (carpel) tissue, such as the outer skin of a peach.

Floret. A small individual flower in a tight cluster of flowers, often associated with sunflowers (Asteraceae).

Fruit. The modified ovary of a successfully pollinated and fertilized flower. The fruit contains the seed(s) within and serves to protect and/or disperse the seeds. By definition, only flowering plants are capable of producing fruits. However, some other groups of plants have evolved fruitlike structures (for example, the conifers yew and ginkgo).

Gall. An abnormal outgrowth, most often found on leaves or stems, caused by several groups of arthropods (including mites, flies, and wasps), fungi, bacteria, viruses, and nematodes.

Gametophyte. The haploid life stage of plants that produce the sperm and eggs. In ferns and lycophytes these are free-living, independent organisms. In the seed plants, they are reduced structures confined within walls of maternal tissue to produce the pollen or ovules.

Geotropic. The directional movement of a structure with or against gravity.

Glume. The outer, papery, leaflike structures surrounding grass flowers, outside of the lemma and palea.

Guttation. The process of exuding excess water from leaves or other structures through pores in the surface. This generally occurs in the early morning.

Haploid. The state of having one copy of each chromosome. Haploid cells can be produced from diploid cells through meiosis.

Haustorium. The branched, tubelike structure in some parasitic plants that penetrates the tissue of a host plant in order to absorb water and nutrients.

Herbaceous. A plant that does not produce a significant aboveground woody stem (that is, not a tree or shrub). Herbaceous plants may be annual or perennial, but they generally die completely or down to their belowground structures every year.

Heterophylly. The phenomenon of one individual plant producing leaves of different shapes and sizes, sometimes varying with age or growth stage, and sometimes in response to the local environment of each leaf.

Heterotroph. An organism that cannot make its own food and must obtain it from the outside world through consumption (for example, most animals).

Hydroids. The water-conducting cells of some mosses.

Hypanthium. In some flowering plants, the portion of the base of the flower where sepals, petals, and stamens fuse, often forming a bowl or cup shape.

Hypocotyl. The region of the plant where the stem system meets with the root system.

Hysteranthy (hysteranthous). The process of a plant flowering before leafing out.

Inferior ovary. An ovary (immature fruit) that sits below the rest of the floral organs (petals, calyx, stamens).

Inflorescence. A cluster of flowers along an axis.

Internode. The span of stem in between nodes.

K-selected. A reproductive strategy characterized by investing more resources into fewer offspring, increasing the likelihood that any individual offspring survives to adulthood. In simple terms, this strategy favors quality over quantity.

Lamellae. Columns of cells filled with chloroplasts that arise from the leaf surface in some species of moss.

Lamina. The broad, flattened portion of a leaf, often serving the primary role of capturing light and performing photosynthesis.

Leaf primordium (*plural primordia*). A small cluster of cells in the shoot apical meristem (SAM) that will go on to develop into a leaf. This is the earliest stage at which plant hormonal signals define the location of a new leaf.

Leaf scar. The mark left behind on the stem of a perennial plant after a leaf falls off. Leaf scars often have shapes and patterns that are unique to a species or closely related group and can be very helpful for identification.

Lemma. The papery portion of an individual grass flower that sits inner to the glume but outer to the palea.

Lenticel. A pore in the bark of a tree that allows for the exchange of gasses with the atmosphere, especially the uptake of oxygen and release of carbon dioxide. Lenticels are especially visible in birches and cherries.

Leptoids. The sugar-conducting cells of some mosses and their relatives.

Lodicule. The modified sepal and petals of a grass flower.

Lower epidermis. The outermost layer of cells on the underside of a leaf, typically containing most of the stomata.

Lycophyte. A clade of vascular plants that disperse using spores and produce single-veined leaves that are often small. Lycophytes first appear in the fossil record about 420 million years ago and are the closest living relatives to the rest of the vascular plants (ferns and seed plants).

Magnoliids. The group of flowering plants that includes magnolias, pawpaws, and cinnamon. This group includes the closest-living relatives of all the rest of the flowering plants.

Marcescence. The pattern in some deciduous trees of holding on to dead leaves for a long period of time before finally dropping them.

Meiosis. A reductive cell-division process leading to the production of four cells that have half of the genetic material of their parent cell. This is the process that produces eggs and sperm in animals, and spores in plants.

Meristematic. A region of stem cells that has the ability to divide and produce new cells that can become any other type of cell.

Mesocarp. The middle layer of fruit (carpel) tissue, such as the juicy flesh of a peach.

Monocot. The evolutionarily distinct group of flowering plants that includes the grasses, lilies, palms, and their relatives.

Monoecious. Plants that bear separate pollen-producing and ovule-producing flowers on a single individual. This is in contrast with dioecious plants.

Multiple fruit. A fruit consisting of many fused carpels derived from many separate flowers (for example, pineapple).

Naked bud. A terminal or axial bud on a tree or shrub that lacks bud scales or stipules.

Node. A point along a stem where a leaf, bud, or branch can arise.

Organogenesis. The process of producing new organs, which are coordinated tissue systems that work together to perform a specific function (or set of functions). For example, a leaf is an organ that performs the function of converting solar energy into sugars in most plants.

Ovary. The structure of the flower that holds the seeds. After fertilization, this ripens into the fruit.

Ovule. The egg-bearing structure of seed plants. It contains the female gametophyte, which houses the egg. The ovule will eventually turn into the seed after fertilization occurs.

Palea. The inner papery portion of an individual grass flower surrounding the lodicules.

Palisade layer. The layer of cells just below the upper epidermis of a leaf. These are usually elongate or columnar, with many chloroplasts. This layer can be one to several cells thick. Its main function is to capture light as it passes through a leaf.

Pappus. A modified, hairy calyx in plants like dandelion, which aids in wind dispersal.

Pentapetalae. The clade of flowering plants that includes roses, mints, asters, and their relatives, all of which have five petals.

Perennial. A plant that lives for more than two years.

Perfect flower. A flower that contains both pollen-producing parts (stamens) and ovule-producing parts (carpels).

Petal. The floral structure immediately inside of the sepals and outside of the stamens and carpels. In many species of flowering plant, petals are specialized in shape, color, and scent to attract animal pollinators.

Petiole. The stalk holding up a leaf.

Phenotypic plasticity. Variation between individuals in one or more characteristics that is explained by the developmental response to different environments, not by genetic differences.

Phloem. The tissue system responsible for moving sugars (produced by photosynthesis or mobilized from long-term storage) throughout the plant body.

Photosynthesis. The chemical reaction performed by plants that uses energy from sunlight to convert water and carbon dioxide into sugar and oxygen.

Phyllotaxy. The spatial arrangement of leaves along a plant's stem.

Phytomer. The repeating unit of the plant shoot system (its aboveground body). A phytomer consists of a leaf, a bud directly above the leaf, and a length of stem. Plants are built of repeating phytomers (leaf—bud—stem—leaf—bud—stem—and so on). Phytomers can be stretched, compressed, or rotated, and their buds can give rise to new phytomers (a branch) or other structures, like flowers.

Pinna. The leaflet of a dissected fern leaf.

Pinnule. The leaflet of a pinna of a highly dissected fern leaf.

Pith. The central region of a stem that develops into ground tissue, not vascular tissue.

Placentation. The way that ovules or seeds are attached to a fruit.

Pollen. The male reproductive unit of seed plants. It is a microscopic grain, often 6,000 times smaller than a grain of rice. Within the pollen grain is the male gametophyte, which contains the sperm.

Prickle. A sharp projection on a plant that develops as an outgrowth of the outermost layer of tissue, the epidermis. Examples can be found on roses and raspberries (yes, they have prickles, not thorns).

Propagules. A term used to define the dispersal unit of a sessile organism.

Proteranthy (proteranthous). The process of producing flowers before leafing out.

Ray floret. The outermost flowers of a sunflower (Asteraceae), generally with showy petals for attracting pollinators.

Rhizoid. A short, threadlike projection growing from the base of some plants like mosses and liverworts. Rhizoids serve a similar function to roots, securing the plant to their substrate and assisting in the uptake of water and dissolved nutrients.

Rhizome. A stem system that typically grows horizontally and underground. However, in some plants, notably many ferns, rhizomes can grow aboveground and may even climb vertically up tree trunks and other structures.

Root apical meristem (RAM). The root equivalent to the shoot apical meristem. It is a zone of stem cells that give rise to all of the cells within a root.

Root cap. The set of cells developed at the very tip of the growing root that helps protect the root as it navigates through the soil.

Root hairs. The epidermal outgrowths of a root that increase surface area and help facilitate the uptake of water and nutrients.

Root system. The counterpart to the shoot system. The embryonic root and all subsequent roots that are derived from it.

R-selected. A reproductive strategy characterized by producing large numbers of offspring while investing very few resources into each offspring. In simple terms, this strategy favors quantity over quality.

Samara. A dry fruit containing a single seed that has a papery winglike tissue projecting from one or more sides (for example, maple, ash, and elm fruits).

Seed. A plant's embryo wrapped in protective tissue (the seed coat) and provisioned with nutrients to grow and survive.

Sepal. The outermost structures of a flower, which enclose and protect the developing flower bud. Sepals are often green and flat, more closely resembling leaves than the other components of the flower. However, in some species sepals take on more elaborate shapes and structures, serving an attractive function.

Sessile. Describing an organism that cannot move from place to place. Plants are sessile, as are fungi and some animals, like sponges and corals.

Shoot apical meristem (SAM). The region of stem cells at the growing tip of a plant.

Shoot system. The generally aboveground plant system that includes the shoot apical meristem that gives rise to subsequent stems, leaves, fruits, and flowers.

Sori (*singular* sorus). Clusters of sporangia on fern leaves.

Spine. A sharp projection on a plant that is derived from an evolutionarily modified leaf. These tend to be organized on a stem in a more regular pattern than prickles or thorns. Examples can be found on cacti.

Spongy layer. The layer of cells just inside of the epidermis on the bottom of a leaf. Cells in the spongy layer are separated by ample air spaces, allowing for gas exchange (carbon dioxide in for use in photosynthesis, and oxygen out as a byproduct of photosynthesis) with the surrounding air via the stomata in the adjacent layer of epidermis.

Sporangium (*plural* sporangia). The site of meiosis and spore development in all land plants.

Sporophyte. The diploid life stage of plants. This is the individual that produces spores through meiosis. In a tree, fern, or herb, this is essentially everything we see in the field, whereas in a moss, the sporophyte is just the small brownish structure that emerges at the tip to accomplish spore production.

Stamen. The structure within a flower where pollen is produced. Stamens often consist of an anther, which houses the pollen, and a filament that holds up the anther.

Stigma. The organ of a flower that sits atop the style and is receptive to pollen.

Stipule. A portion of tissues (often green and leafy) at the base of a leaf where it connects to the stem.

Stomata (*singular* stomate). The microscopic pores in leaves and other plant organs that facilitate gas exchange. Carbon dioxide enters plant tissues through stomata while water escapes. Stomata can open and close in response to many different signals, including drought, sunlight, and hormone levels.

Strobilus (*plural* strobili). A reproductive structure consisting of a stem and modified leaves bearing sporangia.

Superior ovary. A fruit (immature ovary) that sits above the rest of the floral organs (calyx, petals, stamens).

Synanthy (synanthous). The process of leafing out and flowering simultaneously.

Tendril. A modified stem, petiole, or leaf that helps a vine grasp onto a supporting structure. Tendrils exhibit touch-responsive growth, or thigmotropism.

Tepal. The term used to describe sepals and petals together in species where these whorls of the flower are nearly identical and distinguished only by their relative position (for example, tulips).

Thigmotropism. Coordinated growth in response to touch, for example, in the twining stems and tendrils of vines, which grow to wrap tightly around structures they touch.

Thorn. A sharp projection on a plant that is derived from an evolutionarily modified stem. Thorns may be branching, as in the honey locust (*Gleditsia triacanthos*), which reflects their developmental identity as modified branch systems.

Tracheary elements. The cells in the xylem that are responsible for moving water throughout the plant's body. These cells are elongated into tubes and dead at maturity.

Transpiration. The process of water evaporating from the air spaces inside plant tissue (usually leaves).

Trichome. A plant hair. Trichomes may be branched or unbranched, and they may consist of only one cell or multiple cells. Some trichomes have bulbous glands at their tips that secrete sticky biochemical mixtures.

Upper epidermis. The outermost layer of cells on the top side of a leaf (see epidermis).

Vacuole. A large, water-filled sac inside of all plant cells that helps maintain a cell's rigidity. The vacuole may also be used to store chemical compounds within a cell.

Vascular tissues. The xylem and phloem, or the set of tubes that move water and nutrients through the plant body.

Viridiplantae. A clade containing green algae and all land plants, which evolved from within the broader diversity of green algae.

Volatile organic compounds (VOCs). The carbon-based chemicals that usually serve an attractive or defensive function and often have strong smells and/or tastes to human senses.

Xylem. The tissue system associated with water movement in the plant body. Not all cells of the xylem are explicitly for water movement. There are three main cell types produced: the water-conducting cells (tracheary elements), rays or storage/transport cells (parenchyma), and support cells or fibers (sclerenchyma).

Zone of differentiation. The region of the root tip where derivatives of the stem cells differentiate.

Zone of elongation. The region of the root tip where cells expand.

Zygomorphic. Used to describe a bilaterally symmetrical flower (a flower with only one plane of symmetry). Contrast with actinomorphic.

Zygote. The product of fertilization between sperm and egg, producing a diploid cell. This will eventually undergo cell division to develop into an embryo.

Acknowledgments

This book is the culmination of five years of promoting inquisitiveness and enthusiasm for plants through Let's Botanize Inc. During that time, we have benefited immensely from the guidance and support of a number of people and institutions, and we will do our best to highlight many of them here.

As we worked on writing our first book together, we quickly learned that it takes more than two people to make it great. Thank you to the whole Storey Publishing team who brought their brilliance to bear on this project, especially our editor Hannah Fries, copy editor Rowan Sharp, and creative director Carolyn Eckert. During the process of developing the idea and connecting with a publisher, we also received help from Jessica Walliser, David Meerman Scott, Evan Schnittman, Matt Holt, and George Gibson.

Let's Botanize started as a two-person operation but now benefits from the insight and encouragement of an accomplished board of directors, including Sylvia Kinosian, Ash Heim, and Eric Ossola. We would also like to give a shout-out to our inaugural interns, Cruz Gouveia and Lydia Uptain. We have received crucial administrative and legal assistance from Howie Miller, Jason Finger, Robert Shmalo, Barbara Wahl-Ossola, Chuck Ossola, Brian Stevens, Karolyn Richter, Megan Rzonca, and Amy McFarland.

Our work together was born out of the friendship that we cultivated in graduate school in the Organismic and Evolutionary Biology Department at Harvard University. We want to thank our PhD advisers William "Ned" Friedman (adviser to Jacob) and Robin Hopkins (adviser to Ben). Working to build Let's Botanize could have been seen as a waste of time from the perspective of an academic adviser, but their generosity, care, and support of our extracurricular interest in science communication helped grow the roots of our organization with more confidence than we would have otherwise felt. During those early days, we were based at the Arnold Arboretum of Harvard University. This "living tree museum" was an invaluable sandbox for developing our ideas. We want to thank the curatorial and horticultural staff, particularly Michael Dosmann, for maintaining a thriving living collection. We would also like to thank Amy Heuer and Stephen Hill at the Arnold Arboretum and Arielle Moon at the Harvard Museums of Science and Culture for being early champions for Let's Botanize.

We have many friends, family, and colleagues to thank for their unwavering encouragement, including Mindy Cohen, Michel Suissa, Lisa Sherman, Barbara Wahl-Ossola, Alex Ossola, Mark and Joung Goulet, David and Yukari Scott, Andrew Goulet, Greta Wong, Molly

Edwards, Austin Garner, Grace Burgin, Jess Gersony, Min Ya, Kristel Schoonderwoerd, Ellie Mendelson, Sam Church, Morgan Furze, Dylan Wainwright, Andrea Berardi, Tony Raffa, Sam Wooster, Gina Danza, Isaiah Scott, Fay-Wei Li, Pam Diggle, Cynthia Jones, and Rosetta Elkin. Rather than laugh at our earliest attempts at educational media, you told us to keep going and supported us in many ways. Thank you a second time to Andrew Goulet, who provided the spark of an idea that turned into this book. And we would like to sincerely thank our patient and supportive partners, Allison Goulet-Scott and Júlia García Güell, who sometimes see us less often than we see each other.

We would like to thank the individuals who are curious enough about plants to have stumbled upon Let's Botanize and trust us as two of their botanical educators. You are the reason we do this work and wrote this book. Finally, the plants are the organisms that are doing the heavy lifting. It is their diversity of form, function, and evolutionary history that yield such lively stories to share. We are merely their vectors.

Index

Additional Interior Photograpy Credits